Adaptive Multi-reservoir-based Flood Control and Management for the Yellow River

Towards a Next Generation Software System

By

Shengyang LI

Adaptive Multi-reservoir-based Flood Control and Management for the Yellow River

Towards a Next Generation Software System

DISSERTATION

Submitted in fulfillment of the requirements
of the Board for Doctorates of Delft University of Technology
and of the Academic Board of the UNESCO-IHE Institute for Water Education
for the Degree of DOCTOR
to be defended in public
on Friday, May 3, 2013 at 12:30 hours
in Delft, The Netherlands

by

Shengyang LI

born in Shandong, China

Master of Science in Hydroinformatics (with distinction), UNESCO-IHE, Delft, The Netherlands
Master of Engineering in Hydroinformatics, UNESCO-IHE, Delft, The Netherlands
Bachelor of Engineering in Hydrology and Water Resources Management, Chengdu University of
Science and Technology (now Sichuan University), Chengdu, China

This dissertation has been approved by the supervisors
Prof. dr. ir. A. E. Mynett
Dr. ir. I. Popescu

Members of the Awarding Committee:

Chairman	Rector Magnificus, Delft University of Technology
Vice-Chairman	Rector, UNESCO-IHE
Prof. dr. ir. A. E. Mynett	UNESCO-IHE / Delft University of Technology, supervisor
Dr. ir. I. Popescu	UNESCO-IHE, co-supervisor
Prof. dr. ir. G. S. Stelling	Delft University of Technology
Prof. dr. ir. P. van der Zaag	UNESCO-IHE
Prof. dr. I. D. Cluckie	Swansea University, United Kingdom
Prof. dr. Z. Xu	Beijing Normal University, China
Prof. dr. ir. M. J. F. Stive	Delft University of Technology (reserve member)

CSC Press/Balkema is an imprint of the Taylor & Francis Group, an informa business

Published by:
CRC Press/Balkema
PO Box 11320, 2301 EH Leiden, The Netherlands
e-mail:Pub.NL@taylorandfrancis.com
www.crcpress.com – www.taylorandfrancis.com

ISBN: 978-1-138-00102-2

Summary

The Yellow River is known as the 'mother river' of the Chinese people, but is also said to be 'China's Sorrow' or the 'Scourge of the Sons of Han', because in history multiple major floods have had catastrophic effects on people and land along the 5000 km long river reaching from the Himalaya's to the Bohai Sea. Three sections can be distinguished: the upper, middle and lower part of the Yellow River. The upper part is not considered in this thesis. In the middle Yellow River, the Yellow River Conservancy Commission built a number of reservoirs from the 1950's onward, mainly for flood protection. The lower part of the river consists of an 800 km long so-called 'hanging river' where, due to sedimentation, the riverbed exceeds the floodplain level, which makes the floodplain area extremely vulnerable to flooding, in particular in case of high river discharges. In that sense, the lower part of the Yellow River is very similar to some of the polder systems in the Netherlands.

Hence, proper operation of the multi-reservoir-based Yellow River system plays an essential role in regulating floods and minimizing possible damage. Since hydrologic conditions are susceptible to climate change, and indeed the entire basin has been affected by human developments during the past decades, the concept of adaptive management is being explored in this thesis to deal with ever changing reservoir storage conditions and varying river flood conveyance. At the Yellow River Conservancy Commission (YRCC) of the Ministry of Water Resources, China, the decision making process conventionally takes place in collective management meetings.

The general process is that the latest information on hydrological forecasts and reservoir conditions are collected and discussed in a high-level meeting, where the overall goals are set on how to regulate the coming flood. With the existing decision support tool, the effects of operational alternatives are categorized: the less important issues will be dealt with by the technical experts, the more crucial decisions are discussed and agreed upon collectively in the decision making meeting.

In this research, studies were carried out on how to design and implement an adaptive management software tool that concurs with the practice and experience of the YRCC. User requirements, system architectures, computational functions, supporting tools, and man-machine interaction were all explored. The multi-reservoir regulation scheme of the Yellow River requires not only uncontrolled but also controlled release options, including constant release schemes in some particular zones. Moreover, different zones may have quite different storage-release relationships, which require special attention and robust routing techniques. It is also important to deal with the delicate transitions between the various release schemes. A robust algorithm was developed as presented in this thesis.

The Delft numerical flood simulation models were used as the basis for evaluating operational alternatives for the Yellow River. Both state-of-the-art simulation tools from Deltares Software Center, in particular SOBEK 1D2D, as well as the new D-Flow Flexible Mesh version, were applied in this thesis. The two models clearly show the capabilities of present day advanced computer software packages for evaluating alternatives that minimize the flooding extent. Although more work is needed to develop these simulation models into well-calibrated real-time operational tools with much smaller grid size, the potential of the concept is shown in this thesis.

In reservoir operation, traditional methods usually assume the storage-release function to have a simple single-valued relationship, which generally implies uncontrolled release. However, in real-time flood control management, both controlled and uncontrolled release schemes are required. A common feature for flood control and management in the Yellow River often is a combination of uncontrolled and controlled release, which implies a rather complex relation between storage and release.

A robust calculation method was developed in this thesis to deal with 'multi-valued' storage-release relationships. The applicability of the method is generic, in that the release quantity and water level only need to have some generic mathematical relation, and need not be limited to a single-valued function relation.

In an effort to design a good decision support system, different software tools were studied in this research. It was commonly found that different applications require different architectures. Model-centered IPO (Input-Process-Output) architectures are very popular with model-based simulation systems. However, nowadays a data-centered approach is often used as the preferred architecture for decision support systems.

From a functional decomposition point of view, multi-layered architectures are used to create the functions that serve higher level applications. Introducing the concept of a 'real-time-reactive' mechanism as developed in this thesis, has shown to be a good choice for developing such systems. In this approach the model/simulation component is embedded in a user-driven operational loop rather than following a program-driven input approach. This enables a highly efficient and intuitive approach that is very much in line with the actual decision support process at, in this case, the Yellow River. A prototype software tool was developed to demonstrate the proposed solution.

Case study applications show that the proposed approach is feasible for adaptive management and decision support systems. Adaptive real-time reactive management and operation can provide direct and efficient ways to explore alternatives and more efficiently support the decision making process. The tools developed in this research are mainly used to show the outline of a next generation software architecture, which very much favours the 3Di system presently being developed in the Netherlands.

Samenvatting

De Gele Rivier wordt ook wel de 'moeder rivier van het Chinese volk' genoemd, maar staat ook bekend als 'China's verdriet' of 'de gesel voor de zonen van Han', aangezien in de geschiedenis meerdere rampzalige overstromingen zijn voorgekomen langs de 5000 km lange rivier die begint in de Himalaya's en eindigt in de Bohai zee. Er kunnen drie delen worden onderscheiden: de bovenstroom, het midden gedeelte, en de benedenstroom van de Gele Rivier.

Het bovenstroomse gedeelte wordt in dit proefschrift niet beschouwd. In het midden gedeelte heeft de Yellow River Conservancy Commission (YRCC) sinds de jaren 1950 een aantal reservoirs gebouwd, hoofdzakelijk om overstromingen te voorkomen. Het benedenstroomse gedeelte bestaat uit een 800 km lange zogenaamde 'hangende rivier' waarvan, ten gevolge van sedimentatie, de bodem hoger ligt dan de uiterwaarden. Hierdoor is er steeds een kans op overstroming aanwezig. In dat opzicht heeft het benedenstroomse deel van de Gele Rivier veel gelijkenis met boezemwater systemen in de Nederlandse polders.

Vandaar dat het van belang is om de meerdere reservoirs in het Gele Rivier system goed in te zetten om eventuele overstromingen te voorkomen en schade te beperken. Aangezien de hydrologische condities kunnen wijzigen ten gevolge van klimaatverandering en het gehele stroomgebied de afgelopen decennia beinvloed is door menselijke ingrepen, is in dit proefschrift nagegaan of het concept 'adaptief management' hier gebruikt kan worden.

Traditioneel volgt de YRCC een procedure die is gebaseerd op besluitvorming tijdens gezamenlijke vergaderingen. Daarin wordt de laatste informative met betrekking tot de situatie en de voorspellingen bijeengebracht en besproken in een bijeenkomst met hoge afgevaardigden, die de doelen vaststellen en de vereiste middelen toekennen om overstromingen te voorkomen. Met het bestaande instrumentarium worden de gevolgen van operationele alternatieven in categorieen ingedeeld: de minder belangrijke zaken worden gedelegeerd waarna de belangrijke beslissingen gezamenlijk worden genomen. In dit proefschrift is nagegaan hoe software systemen kunnen worden ontworpen en geimplementeerd zodat deze geschikt zijn voor gebruik in de praktijk van de YRCC. Toepassingseisen, system architecturen, rekenfuncties, ondersteunende gereedschappen en de gebruikersinteractie met computersystemen zijn alle onderzocht.

Om de verschillende reservoirs van de Gele Rivier goed te kunnen reguleren, zijn meerdere opties nodig met betrekking tot gecontroleerde en ongecontroleerde uitstroom. In dit proefschrift is een robuste method ontwikkeld die al deze opties aan kan.

De Delftse numerieke overstromingsmodellen hebben als basis gediend voor de verschillende operationele afwegingsmodellen die zijn ontwikkeld voor de Gele Rivier. Zowel de standaardmodellen van Deltares Software Center (hier SOBEK1D2D), als ook het nieuwe D-Flow Flexible Mesh zijn gebruikt in dit proefschrift. Beide modellen laten overtuigend zien welke mogelijkheden er vandaag de dag al zijn om overstromingen te simuleren, en welke in de nabije toekomst mogen worden verwacht. Hoewel er nog het nodige te doen staat voordat deze simulatiemodellen kunnen worden ingezet op veel kleinere rekenroosters en in goed gecalibreerde beslissingsondersteunende systemen, zijn het principe en de mogelijkheden van dit concept in dit proefschrift aangetoond.

Bij reservoirbeheer worden gewoonlijk enkelvoudige functionele relaties gebruikt tussen instroom, opslag en uitstroom. Echter, bij het beheer van meerdere reservoirs kunnen deze relaties ingewikkelder worden, zoals het geval is voor de Gele Rivier. In dit proefschrift is een robuste methode ontwikkeld die om kan gaan met meerwaardige relaties tussen opslag en uitstroom en niet beperkt is tot enkelvoudige functies.

Verschillende software methoden en systemen zijn onderzocht om te komen tot een goed beslissingsondersteunend systeem. Daarbij bleek dat verschillende toepassingen om steeds verschillende ontwerpen vroegen. Het klassieke Input-Proces-Output ontwerp is zeer geschikt voor numerieke simulatiemodellen. Echter, voor beslissingsondersteunende systemen lijkt een data-georienteerde aanpak te prefereren.

Uit het oogpunt van functionele decompositie kan een meerlaagse software architectuur worden gebruikt om specifieke functies te creeren die op een hoger niveau kunnen worden toegepast.

In dit proefschrift is het concept van een 'real-time-reactive' system ontwikkeld wat een goede keuze bleek. In deze benadering worden de simulatiemodellen ingebed in een gebruikers-georienteerde operationele toepassing en niet zozeer in een omgeving voor het draaien van simulatieprogramma's. Daarmee wordt een efficiente en intuitieve omgeving gerealiseerd die goed aansluit bij het daadwerkelijke process van besluitvorming bij de Gele Rivier. Het principe hiervan is als prototype ontwikkeld en aan gebruikers van de YRCC gedemonstreerd.

Op basis van een aantal cases die zijn ontwikkeld kan worden geconstateerd dat de voorgestane aanpak geschikt is voor met name adaptieve systemen. Nieuwe informatie kan direct in het besluitvormingsproces worden ingevoerd zodat nieuwe alternatieven kunnen worden ontwikkeld en beslissingen kunnen worden genomen. De technieken die in dit proefschrift zijn onderzocht schetsen de contouren van een volgende generatie software systemen, zoals het 3Di system dat momenteel in Nederland wordt ontwikkeld.

中文概要（Summary in Chinese）

黄河是中华民族的"母亲河"，她孕育了灿烂的中华文明。然而，黄河又由于历史上频繁的决口泛滥和洪灾而被称作"心腹之患"。黄河通常被分为上游、中游和下游三个段，其中上游河段不在本研究覆盖范围之内。自上个世纪五十年代以来，多个以防洪为主的大型水库在中游相继建成。黄河下游段长度近 800 公里，由于长期的泥沙淤积而形成了所谓的"地上悬河"，主槽河床高程高于滩地，这使得这些滩地极易受到洪水的危害，如遇大洪水，滩地易遭受大的灾害，这与荷兰的低洼垦地（polder）所遇到的情况很相似。

因此，中游水库群的联合调度对于调节中上游来水并减少下游潜在的灾害起着关键的作用。由于气候变化和人类活动的影响，适应性管理对于实时控制和管理洪水具有重要意义。在黄河防洪的管理机关黄河水利委员会，防洪管理的决策通常是以一种集体决策会议的形式进行。

在此会议上，防洪决策的一般过程是，首先获取并讨论最新的雨水工情信息，然后确定洪水的调度目标和调度方式。在现有决策支持工具的支持下可以对调度方案的实施效果进行分析和评估。对于次要的问题将很快达成一致意见，而更重要的问题将通过集体的分析和决策来做出决定。

本研究旨在从黄河水利委员会的实际需求和应用出发，研究如何设计和开发用于适应性管理的决策支持软件工具。围绕这一研究目的，用户需求、系统体系构架、计算功能模块与系统功能构成、决策支持与人机交互方式等将被重点分析和研究。从洪水调节计算而言，黄河中游水库群联合防洪运用不仅涉及到固定开启方式（非控制性）泄流，而且也有非固定开启方式（控制性）泄流，包括控制下泄流量或流量过程、控制某一水位等。此外，水库还有不同调度规则与方式下的不同调度运用水位区段，不同区段有不同规则下的水位（蓄量）与泄流关系，相较于普通的水库调节（调洪）演算，这里需要更为可靠而精确的计算方法。同时，不同调度（尤其是控制性泄流）方式下调度方式之间的转换与过度也需要精细的算法以实现正确的水量平衡与泄流过程计算。本研究提出了一个新的水库调节计算方法并提出了不同调度方式之间转换的精确算法。

在本研究中，荷兰代尔夫特水力学研究院开发的洪水模拟软件工具被用来对水库调度方案所涉及的黄河下游可能的洪水淹没进行分析与评估。该研究院软件中心所开发的 SOBEK 1D2D 统一网格和 D-FLOW 非统一网格模拟工具都被应用到本研究中。两个模型工具都体现了当今洪水淹没模拟技术的最先进水平。尽管要真正应用到生产实践中的洪水淹没分析，尚需要使用更小的网格和开展更多的率定工作，但是，上述模拟工具的基本应用过程和方式已经在本研究的应用中被概要地体现出来。

在洪水调节计算时，传统方法通常基于水位（或蓄量）与泄流间呈现单值函数关系来进行，这通常意味着固定开启方式的非控制性泄流，然而在防洪调度和管理中，控制性和非控制性泄流都存在。黄河防洪调度中一个常见的方式就是控制性和非控制性泄流共存和混合进行，这种情况涉及到复杂的水位（蓄量）泄流关系，并使得精确的水流调节计算更为复杂化。

本研究中提出了一个新的、可实现稳定可靠计算的水库调节计算方法（水库调洪演算方法）。该方法可以对具有多值关系的水位（蓄量）泄流关系进行调节计算。这一方法具有简单、通用、精确的特点，它仅需要水位（蓄量）泄流之间具有数学概念上的关系（relation）关系，而非限于单值的函数关系。

从开发先进实用的决策支持系统出发，本研究中对不同的软件工具及相关软件构架、运行机制进行了研究和分析。不同的应用软件需要具不同的、带有针对性的软件构架和运行机制。就结构而言，以模型为中心（model-centred）的IPO（输入-处理-输出）简单构架在模型分析工具软件中比较常见。而今以数据为中心（data-centred）的构架也已被提出来，在决策支持系统设计中也有应用。

在计算功能构成方面，本研究提出用一个多层体系结构用于逻辑地构筑系统不同层级的功能模块，这些模块以为其更上一层提供应用服务和构成其功能组件为主要设计原则。同时，从系统整体构架形成的角度出发，为构建下一代软件系统工具，本研究提出将"即时响应"（real-time reactive）机制应用于决策支持系统构架设计和系统运行方式中。在这个即时响应机制中，模型/模拟作为基本功能组件之一被纳入到一个用户操作需求驱动下的运行循环之中，并作为其环节之一参与运行，而不像一般模拟系统那样处于运行和操作的中心。在研究中，基于上述思路开发了基于即时响应机制的黄河水库群防洪调度决策支持软件工具以用于实现和体现上述机制与设计理念。工具中包括所有必要计算模块和必要的、能体现即时响应机制的人机交互用户界面，并很好地实现了系统整体构架与操作运行机制的有机结合。

实例应用证明本研究所提出的系统开发方案适用于开发适应性的决策支持系统。这样的系统可以在防洪调度决策过程中对决策过程提供更好的支持，为防洪方案分析和优化提供更为直接、更为高效、更为快速的支持手段。本研究中提出的下一代软件系统的体系构架与设计思路，与由荷兰多个水利管理、水力学和软件工具研发等部门共同开展的、面向未来河流系统管理的"3Di"项目理念有着诸多相似之处。

Table of Contents

1 Introduction 1

  1.1   Background 1

  1.2   Problem description 5

  1.3   Research questions 6

  1.4   Objectives 7

  1.5   Outline of the thesis 8

2 Yellow River flood control, management & decision support 9

  2.1   Yellow River general description 9

  2.2   Yellow River flood control engineering system 12

  2.3   Yellow River flood control management and operational rules 16

    2.3.1   *General aspects of flood control management rules on the mid-lower Yellow River ... 17*

    2.3.2   *Flood control computational scheme 21*

  2.4   International developments in flood management 24

    2.4.1   *The International Flood Initiative 24*

    2.4.2   *The EU Flood Directive 26*

    2.4.3   *Flood Control 2015 Program in The Netherlands 26*

    2.4.4   *Flood Risk Management Consortium in the UK 27*

  2.5   Software systems for operational management 28

  2.6   The need to improve decision support tools 30

    2.6.1   *Adaptive management 30*

    2.6.2   *Adaptive decision support software architectures 33*

    2.6.3   *Internal execution pattern and operational mechanism 34*

    2.6.4   *Flexibility on function compositions and rule management 34*

    2.6.5   *Evaluation of alternatives for flooding simulation 35*

    2.6.6   *Other requirements for improvement 36*

  2.7   Summary 37

3 Multi-reservoir-based flood regulation 39

  3.1   Basic reservoir regulation equations 39

  3.2   Yellow River multi-reservoir flood regulation 42

  3.3   Review of multi-reservoir simulation tools 45

  3.4   Calculations required for YR reservoir operation 48

    3.4.1   *Reservoir routing conditions and routing methods 49*

    3.4.2   *Routing method requirements for fixed H-q relations ... 54*

    3.4.3   *Requirements for controlled release routing 55*

  3.5   Reservoir operation and downstream safety 59

  3.6   Summary 63

4   Numerical flood simulation on lower Yellow River 65

4.1   Introduction 65
4.2   Numerical modeling of flood events 68
4.3   The Lower Yellow River flood computational domain 70
4.4   Numerical modeling of YR using a uniform grid (SOBEK) 75
4.5   The 2D flood model using flexible grids 83
    4.5.1   Model setup of D-Flow FM *83*
    4.5.2   Simulation results *85*
4.6   Some relevant aspects in real-time operation 89
4.7   Summary 91

5   A flexible reservoir routing method 93

5.1   Introduction 93
5.2   Background 94
    5.2.1   The governing equations for routing calculations *94*
    5.2.2   Variables of the reservoir routing equations *95*
5.3   The reservoir level-storage-release relations 97
5.4   A new routing method: the Cross Line Method (CLM) 99
5.5   Step by step algorithm of the proposed CLM method 103
5.6   Discussion 106
    5.6.1   The flexibility of the Cross Line Method *106*
    5.6.2   CLM application for special reservoir storage conditions ... *109*
5.7   Conclusions 111

6   Next generation software architecture for decision support ..113

6.1   Introduction 113
    6.1.1   Adaptive management application *113*
    6.1.2   Yellow River approach *114*
6.2   Functional decomposition of a DSS system for multi-reservoir-based regulation 115
    6.2.1   Next generation software environment for river applications ... *115*
    6.2.2   The functional structure and basic components *117*
    6.2.3   River flow routing due to multi-reservoir regulation *120*
6.3   Decision making process in flood control and management 123
6.4   User requirements on a DSS for flood control and management 126
    6.4.1   User requirements in terms of operational details *126*
    6.4.2   System architecture and execution procedure *127*
    6.4.3   User interfaces and graphical display *128*
    6.4.4   Use of IT resources and tools *128*
6.5   A new approach in developing better DSS 129
    6.5.1   The real-time reactive mechanism for DSS *129*
    6.5.2   End-user based graphical interfaces for decision making ... *131*
    6.5.3   Exchange of ideas with relevant users *139*

6.6   Computational technologies for next generation software systems 139

    6.6.1   Supercomputing *139*

    6.6.2   Pre-computed scenarios and historical reference base *139*

    6.6.3   Training ANN on pre-computed results *140*

6.7   The 3Di Water Management project 140

6.8   Conclusions 141

7   Case study applications 143

7.1   Yellow River flood control rules and operational conventions 143

7.2   Case study 1 - historical flood 148

    7.2.1   General information of the 1996 flood *148*

    7.2.2   Reservoir operation in the proposed DSS *151*

    7.2.3   Selection of the operational alternative *159*

7.3   Case study 2 - flood under climate change conditions 160

    7.3.1   General information on the flood *160*

    7.3.2   Preliminary analysis of the flood *163*

    7.3.3   Rule-based alternatives and decision making *166*

7.4   Discussions 169

    7.4.1   The process of obtaining a good solution *169*

    7.4.2   Reservoirs' releases and downstream resultant hydrograph *169*

    7.4.3   Support for decision making *170*

    7.4.4   Other issues *171*

7.5   Summary 172

8   Conclusions and Recommendations 173

8.1   The Yellow River and floods 173

8.2   The operational process and new requirements 173

8.3   Multi-reservoir operations and requirements 175

8.4   Advances in numerical flood simulation 176

8.5   A flexible method for reservoir routing 177

8.6   Software architecture for flexible decision support 178

8.7   Case studies for the mid-lower Yellow River basin 180

8.8   Recommendations 180

1 Introduction

This chapter introduces the research done during the PhD, the rationale for it, the objectives and the outcomes. The chapter starts with the presentation of the Yellow River flood control and management practices with the main focus on the decision support aspects related to it. Then the research questions are derived and objectives are formulated. The chapter ends with an outline of the thesis including an overview of the structure and content.

1.1 Background

Reservoirs together with river channels constitute a complex network system for storing, regulating and conveying flows. In such systems, the operation of reservoirs is essential for the control and management of floods. Protection of the people and interests in the relevant regions is an important issue as well. In a river basin with serious flood threat, such as the case of the Yellow River (see Fig. 1.1), in China, forecasting, control and management of floods are important issues for the river management authorities. From flood forecasting point of view the decision makers need rainfall-runoff models and a flood forecasting software system. For flood control and management, we need a software system to do all the necessary computation and analysis that will work out good strategies and alternatives to deal with the floods, and eventually to help the decision-makers to achieve good solutions (Cui et al, 1998).

Different rivers have different flood protection and management engineering systems. Moreover the ways to tackle the flood can be different from one river to another. Usually, for large rivers, the built reservoirs on the main river and its tributaries play a dominant role in dealing with big floods. Fig. 1.2 shows the multi-reservoir river system of the mid-lower Yellow River. In some areas, flood detention/retention basins might be used if there are no reservoirs or if reservoir capacities are not sufficient to regulate extreme floods.

1

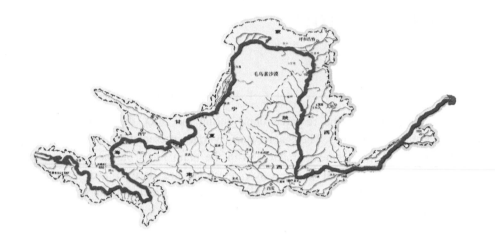

Fig. 1.1 Map of the Yellow River basin
(source: Yellow River Conservancy Commission)

Climate change could in future lead to more severe and extreme flood events (Trenberth K., 2005). According to the 2007 report of the United Nations Intergovernmental Panel on Climate Change (IPCC), "...*the frequency of heavy precipitation events has increased over most land areas*". Available research suggests a significant future increase in heavy rainfall events in many regions of the world, including some in which the mean rainfall is projected to decrease. The resulting increased flood risk poses challenges to society, physical infrastructure and water quality. It is likely that up to 20% of the world population will live in areas where river flood potential would increase significantly by the 2080s (IPCC, 2007). According to predictions for future climate change, the frequency of extreme events will increase in the Yellow River basin and therefore it is a must to enhance the ability to cope with extreme natural events and increase the level of flood control and disaster alleviation under complicated conditions (UNESCO and YRCC, 2011).

Adaptation management to climate change conditions and related decisions have been studied by many researchers in the water resources management and planning phase (Dessai and Hulme, 2007; Dessai et al, 2009). There is a clear need for improving the knowledge about processes involved in adaptation decisions. This knowledge includes information on the steps taken during the decision process, the rationale behind them, handling of uncertainties, choices of adaptation types, conditions and timing, and the consequences of adaptation strategies and/or measures (Smit and Pilifosova, 2001; Burton, 1997) In real-time operation, adaptive flood control and management becomes crucial, since spatial and temporal dynamic rainfall/flood patterns may change between different rain events or even within one event, in particular if there will be more severe conditions induced by climate change. In case of a river with many reservoirs that need to be operated in series or in parallel, there are clearly defined rules for their "collaborative" operation in order to control and manage the floods. However, real-time operation always implies decisions and actions based on real-time situations, which will usually somewhat differ from what have been given in the reservoir operating rule books. Hence, adaptive management is crucial to deal successfully with all kinds of floods.

The YRCC considers floods not only as a threat, but also as an opportunity. The former Commissioner of YRCC Mr. Guoying Li (2005) categorized three types of measures to tackle floods: (1) control floods, (2) manage and make use of floods, and (3) create and make use of artificial floods. To create artificial floods means to make use of the river flow and reservoir storage in order to produce flows which resemble natural floods. These kind of artificial flows are used to transport more sediment or to produce different other benefits for people. The flow and sediment transport coupling regulation (Li, 2009) is an example of such measure. All these actions have put forward more requirements for better efficient supporting tool.

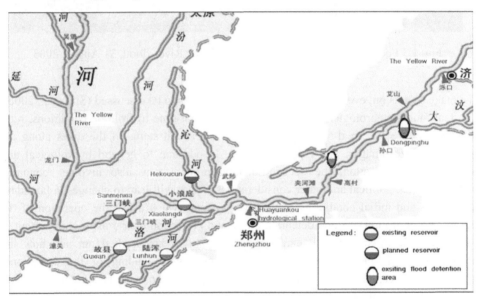

Fig. 1.2 The main flood regulation reservoirs and flood detention areas on the mid-lower Yellow River (source: YRCC)

In China, the flood control decision making process of a big river would probably happen with real-time analysis, collective discussions, developing alternatives, and jointly making decisions by relevant decision makers. Experts and technical people supporting the process are involved, as are staff who are in charge of flood forecasting, including developing real-time operational alternatives (Cui et al, 1998).

Fig. 1.3 shows a YRCC flood control management meeting, which took place during a flood on the middle Yellow River. The meeting was joined by decision makers of YRCC together with experts and relevant heads of different branches in YRCC.

Fig. 1.3 Food control meeting on middle Yellow River flood, 31 August 2003
(source: YRCC)

To make a decision, every relevant aspect of a flood has to be discussed (Shu et al, 2006), i.e. real-time meteorological conditions, forecasts, real-time hydrologic conditions, real-time flood forecasts, real-time flood situation, the actual status of the dikes along the river, hazard estimation and risk assessment. In addition to general hydrological and engineering information, the decision-making process may also involve economic, environmental, political and other considerations. These will lead to changes in boundary conditions and initial conditions of the river system, which affect the operation of the reservoirs, allocation of flood detention areas, and protection measures. It is important to study how flood control and management alternatives are worked out and how the operational decisions are made. Without a good understanding of the operational practices, no accepted system can be constructed, as has been experienced time and again in cases of failure of many software systems.

Since the early 1980s, management authorities of the large rivers in China have made a continuous effort to develop flood control and management software systems. However, end-users of these software systems (including decision makers) are usually not quite satisfied. The requirements for better systems have been recognized in many practical situations, which is why the concept of adaptive management has become an important principle for managing flood problems under changing engineering and climatic

4

conditions. That is why the authorities of these rivers and reservoirs are still working hard to build new software systems.

In the Project Book of National Flood Control and Draught Relief Commanding System (2006), Ministry of Water Resources, China, it was concluded that *"till now, full-function, handy systems for real-time operations are still badly needed by the large river authorities"*. So, the situation of the existing systems, both in YRCC and in other river authorities in China, should be studied carefully to find out things that are not useful to the users and the ones that the users really need.

From solely the computation or processing point of view, a flood control and management software system would include setting up a computational environment, technical processing of flood control rules, reservoir regulation computation, flood detention computation, river flow routing, hazard estimation, risk analysis, data processing and retrieving, graphic presentation, etc. (Li and Mynett, 2009). However, this is not enough for developing a successful operational DSS system.

The challenge for YRCC is to build a system which most efficiently can help the staff in charge of developing alternative scenario's to work out adequate alternatives and help decision makers to achieve good solutions. Any DSS system must be built to be 'powerful-in-function' and 'concise-in-structure', 'user-friendly' and most of all 'efficient and fast', in order to support the decision making process.

1.2 Problem description

Until now, flood control and management support software systems have often failed to fulfill the requirements of the actual operators of such systems. The concept of *adaptive* flood control and management imposes further constraints in order to achieve more flexible and efficient supporting tools. Hence the research towards developing a successful adaptive management tool needs to cover several issues such as: an adaptive approach to flood control and management; building several potential alternative solutions to a given flood problem; evaluating the solutions through a numerical flood simulation model; establishing a favorite architecture for the flood control management and decision support system (DSS); design and implementation of the basic components; developing appropriate user interfaces, man-machine interaction and execution modes; exploring efficient calculation methods.

The research towards an adaptive management and support tool has to follow a systematic approach involving both the high level software architecture as well as basic computational kernels, and man-machine interaction styles. In multi-reservoir operation, the reservoir routing for just one single reservoir flow regulation is done by calculating the mass balance of the reservoirs' in-pool water storage. However, a regulation management and operational oriented DSS will include more elaborate calculations such as; keeping a fix storage level; releasing a constant discharge; releasing a varyiable discharge in time; releasing environmental flows; etc. If the gates are operated during the routing period, (i.e. controlled release operation), then the routing process becomes complex (Chow, 1964). Adaptive management implies the use of sophisticated operational scenarios based on both general operational rules and real-time consideration of all relevant conditions. Hence, the principles of controlled release operation for flood control and management have to be explored in depth.

1.3 Research questions

The research area of this PhD research is the mid-lower Yellow River, which covers the Sanmenxia to Huayuankou sub-catchment (the so-called San-Hua area) and the upper part of lower Yellow River. The four Yellow River reservoirs which are key engineering system for flood control and management of the mid-lower Yellow River, are all located in the San-Hua area. This multi-reservoir system consists of four existing reservoirs, Sanmenxia and Xiaolangdi on the main river reach, and the Guxian and Luhun reservoirs on the Yiluo branch (see Fig. 1.2). The lower section of the Yellow river channel, enclosed by dykes, is the main research area for flooding in this PhD study.

Given the above-mentioned complex system, the following research questions are identified as the main research topics of this thesis:

- What are the general processes of Yellow River flood control, management and what are the requirements for next generation decision support systems?

- What is the operating procedure in multi-reservoir-based Yellow River flood control and what are the requirements for adaptive multi-reservoir-based flood management?

- What are recent developments in numerical flood simulation and can these be used for mid-lower Yellow River inundation prediction?

- Can a method be developed for interactive operation of multi-reservoir operation and how can such method be made robust?

- What software architectures are available from literature and what are the requirements for enabling adaptive management in decision support?

- Can such approach improve the decision making process for the YRCC?

These questions provide the starting point of the research on adaptive flood control and management, including the development of a proposed supporting tool.

1.4 Objectives

The overall objective of the research is to identify the implications of adaptive multi-reservoir-based flood control and management, and to find ways of developing successful decision support tools. The main effort is to identify the requirements for adaptive management and to explore the types of support tools that are necessary to develop a next generation DSS system for adaptive flood control and management.

Following from the overall objective, the research will focus on:

1) The analysis of the general practices in flood control and management commanding activities, especially the flood control and management decision making process; analysis of the main problems to be solved by the flood control decision support systems; definition of the requirements for adaptive management and supporting tools; the analysis of users, including decision makers.

2) Exploring the principles of adaptive management in flood control and management operations which is now facing more uncertainty and extreme condition under climate change impacts and anthropogenic influence; adaptation management as a concept to deal with future extreme situation and events; analysis of adaptive management support tools.

3) Research on new system architectures with adaptive features, internal execution mode and user operating mechanisms for developing a next generation software system, including the interaction between man and machine, especially for decision support; the 'real-time reactive' mechanism will be developed in this part of the research.

4) Research on new and basic methods of routing/regulation calculation that are needed for high-level robust and flexible computational applications will be explored and developed; development of basic computational routines and computational core which can best meet the requirement of a good DSS tool.

5) Development of a prototype system with necessary modules, consisting of reservoir regulation, river flow routing and 2D overland flow simulation models; development of operational prototype functions of a DSS demonstrating the new approach.

6) Applying the approach to historical floods or flood drill data and possible extreme cases that could happen in future due to climate change.

1.5 Outline of the thesis

Chapter 1 introduces the scope of this research. The overall objective for creating an adaptive management environment with good supporting tools is briefly explained. Research questions are formulated and main approaches are presented. The outline of the thesis is given with an overview of content and structure.

Chapter 2 starts with providing the background of the Yellow River flood control and management practices. The engineering systems (reservoirs) constructed for flood control and management of the Yellow River are introduced. The general processes and operating principles of the large river authorities in China are presented and the requirements for creating a computer-based environment for decision support are defined. An overview is given of internationally established software tools for flood control and management.

Chapter 3 shows a simple principle of how synchronised multi-reservoir operation can reduce the flood volume by a factor of two and delay the peak arrival time by hours. The need to develop robust schemes for reservoir operation is elaborated with the aim to deal with typical multi-valued storage-release relationships encountered in the Yellow River.

Chapter 4 focuses on exploring the state-of-the-art in numerical flood simulation. The Delft SOBEK-1D2D and the D-Flow/FlexMesh software modules of Deltares are used to develop a model for the lower Yellow River which is then applied to predict and evaluate various operational alternatives. The simulation area focuses on the upper part of the lower Yellow River (the so-called 'hanging river' section).

Chapter 5 introduces a reservoir routing method especially developed in this research for achieving robust and flexible computational performance. It enables extending single-valued storage-release functions to quite arbitrary multi-valued storage-release relations.

Chapter 6 explores some of the architectures of present-day decision support systems and introduces a new approach. A real-time reactive mechanism is proposed as the basis for a DSS system architecture and a simple example is presented to demonstrate the approach. Preliminary conclusions are drawn on how to develop a next generation decision support system with advanced flood simulation capabilities and rapid user interactions.

Chapter 7 provides a case study application of the proposed approach. Flood data from the YRCC flood control and management drills are used as a basis to simulate real-time operations with special focus on optimizing alternatives.

Chapter 8 summarizes the main findings and provides conclusions and recommendations from this research.

2 Yellow River flood control, management & decision support

This chapter starts with introducing background and engineering systems for the Yellow River flood control and management. The general process regarding the flood control and management activities of Yellow River authorities is studied to see what the requirements for decision support are. The software tools used by YRCC are studied in order to find out the requirements for a better decision support system. These requirements are the driving force for this research.

2.1 Yellow River general description

The Yellow River, originating from the Tibet highlands in West China, at an elevation of 4,500 m.a.s.l, flows through nine provinces and autonomous regions. The river has a total length of 5,464 km, a basin area of 795,000 km², and eventually flows into the Bohai sea. According to its basin size, the Yellow River is the second largest river in China, just next to the Yangtze. The Yellow River is called "the cradle of Chinese civilization" because it was the birthplace of ancient Chinese civilization and the most prosperous region in early Chinese history.

The Yellow River basin is of continental climate. It is located in the semi-humid and semi-arid zone of west and middle China (Xu and Zhang, 2006). The regions above Lanzhou area are the semi-humid zones; just below Lanzhou, the north-west part of the river locates in the arid zone of west China. The catchment area in the south of Xi'an city, the capital of Shaanxi province, together with the area near Mountain Taishan presents features of the humid zone. The Shanxi province and Henan province are semi-humid and semi-arid area (see Fig. 2.1.)

The Yellow River has the biggest natural annual sediment transport amount in the world, 1.6 billion tones, with a natural annual runoff, 58 billion m³. Before the Xiaolangdi dam was put into operation, one quarter of the sediment load would deposit in the lower river channel.

Fig. 2.1. Geographical types of the Yellow River basin (source: YRCC)

The heavy sediment load from the upper resulted in a continuous rise of the river bed and made the lower Yellow River a 'hanging river'. The river has been known as the 'river of hazard' for its frequent embanks failures and heavy flood damages in history. Different reaches of the river have different flood threats, while the lower Yellow River has the biggest and devastating flood problems because of its 'suspended' nature. The main runoff and major transported sediment comes from different areas of the basin, as shown in Fig. 2.2. From the figure, it can be seen that the area above Lanzhou produces 56% of annual runoff, while 90% of the sediment and up to 100% of all coarse sediment, of the Yellow River, yields in the Loess Plateau.

The Yellow River is traditionally divided into three reaches (Fig. 2.3): (i) *the upper reach*, which is above Hekouzhen, (ii) *the middle reach* which is the part between Hekouzhen and Taohuayu (very near to Zhengzhou), and (iii) *the lower reach* which is below Taohuayu. The upper reach is less populated because of its harsh environment. The upper part of the middle reach, where the river goes from north to the south, is the famous Loess Plateau for its fine and easily-eroded soil.

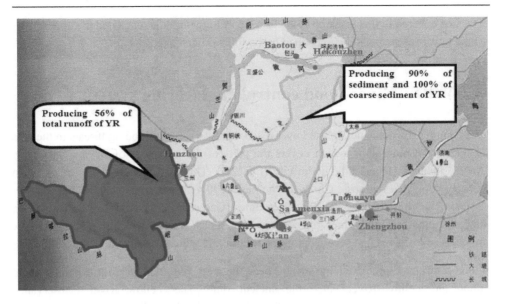

Fig. 2.2. Different source areas for sediment load and runoff in the Yellow River basin
(source: YRCC)

Most of the sediment of the Yellow River originates here and is transported downstream. The lower part of the middle reach, located in central China, is the transition section, the one in which Yellow River flows from mountain areas to the North China plain. The lower Yellow River is the hanging section and the areas outside the dikes (except for the small mountainous area near Mountain Taishan) are no longer part of the catchment area of the Yellow River, because the river becomes an isolated channel. As such, the lower reach, just simply conveys the flow from the middle reach to the Bohai sea.

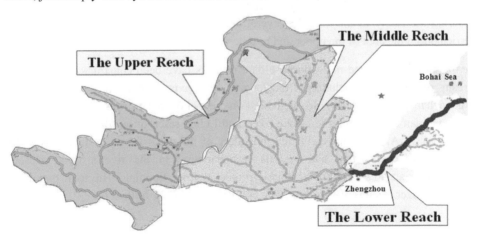

Fig. 2.3. The three reaches of the Yellow River (source: YRCC)

The area covered in this research is the lower part of the middle reach and the upper part of the lower reach as will be explained in the following chapters.

2.2 Yellow River flood control engineering system

The hanging river channel, in the lower reach, leads to serious flooding threats to the area in the north-China plain, which covers the parts of the provinces of Henan, Shandong, Anhui, Jiangsu, Hebei and Tianjin city (Fig. 2.4). This area has always been the most densely populated region in China. In the past 2600 years (before late 1940s), the lower Yellow River has changed its channel 26 times (Gao et al, 1991), and had failures of the embankments for 1590 times (on average two times every three years), causing huge numbers of lost lives and severe damages of farmland with a vast area up to 250,000 km^2.

Flood control and protection has always been the most important task of the people living along the Yellow River since the ancient time. The Yellow River Conservancy Commission (YRCC), a big branch of the Ministry of Water Resources of China, was established in 1946, to manage the Yellow River, with one of its major roles being flood control and management. YRCC also acts as the Yellow River Flood Control and Drought Relief Command Headquarters. In the last sixty years, great effort has been made to enhance the safety of the lower Yellow River (Li, 2005).

Since late 1940s, the downstream dikes have been heightened and strengthened three times. In the same time a preliminary downstream flood control engineering system had been carried out, the so called retaining flood on the upper stream, discharging floods through the downstream channel, dealing with extreme floods with detention and retention areas by both banks". These engineering works have been very successful in ensuring the safety of the dike for several decades. Soil conservation works have been carried out in the area of the upper and middle reaches.

Ten large/medium-sized multi-purpose dam projects have been constructed on the main stream of the river, including Longyangxia, Liujiaxia, Sanmenxia, Xiaolangdi and Wanjiazhai. Many big reservoirs were built on the Yellow River tributaries, as well, such as the Guxian reservoir on the Luohe river, and Luhun reservoir on the Yihe river in the middle reach. All these have ensured the safety of the dikes for more than sixty years, during flood events that had various reappearances in the river's flood seasons. Downstream an area of 120,000 km^2 is protected from flooding. The total population within the river basin is 107 million. The cultivated land is 12 million ha.

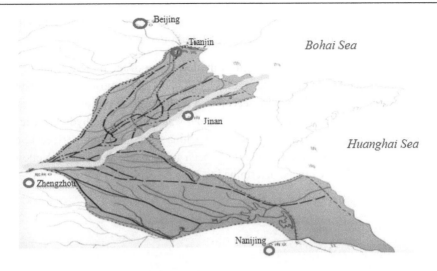

Fig. 2.4. Historical flood extent on the Yellow River in case of dyke failure and channel changes (source: YRCC)

The four reservoirs, Sanmenxia, Xiaolangdi, Guxian and Luhun (Fig. 2.5 and Fig 2.6) play the most important role in flood control and management of the mid-lower Yellow River. Sanmenxia reservoir, which is located at the main stream entrance, can capture all runoff from its upper catchment. From Taoyuayu, a place very near to Huayuankou, (point A, Fig. 2.6) to the river mouth, the Yellow River cannot get runoff contribution from the catchment anymore because it becomes a hanging channel. Therefore, the section between Sanmenxia to Huayuankou (or simply the San-Hua reach or San-Hua sub-catchment) is the most important river-reservoir system for the regulation of floods. These floods are a threat to safety in the lower Yellow River nearby regions. The area of the Sanmenxia - Huayuankou section has a big runoff contribution for the mid-lower Yellow River. This is also a key reason for the construction of the reservoirs in this area.

The two blue areas represented in Fig. 2.6 are the flood detention areas for the mid-lower Yellow River, Beijindi in the upper part that is designed for extreme floods, and Dongpinghu, in the lower part, designed for big and extreme floods. Whether these two areas are put into operation or not, is determined based on the discharge measured in the hanging part of the river. If after the regulation of reservoirs, the flood discharge is still too big, for the hanging river, then decisions will be made whether the flood detention areas will be used or not. The use of these areas is quite rare nowadays, therefore they are not considered in the present research.

Xiaolangdi dam, a key national project, is located 40km north of the city of Luoyang, in central China's Henan Province. The Xiaolangdi reservoir, which went into operation in 1999, is at present, the most powerful flood regulation project for the flood control and management of the middle and lower Yellow River. Its aggregate design storage is 12.65 billion m³ with a flood regulation capacity of 5.0 billion m³.

Xiaolangdi reservoir is located at the exit of the last gorge of the middle reach of Yellow River, 130 km downstream of Sanmenxia reservoir and 128 km upstream of Huayuankou of Zhengzhou city. The Xiaolangdi reservoir is a multi-purpose project for the control of floods, ice jams, siltation, industrial and municipal water supply and hydroelectric power. The region surrounding the lower reaches of China's second longest river is densely populated and a major agricultural area. It has been subjected to devastating Yellow River floods in the past, which China is determined to end.

Fig. 2.5. (i) Sanmenxia reservoir

Fig. 2.5. (iii) Guxian reservoir

Fig. 2.5. (ii) Xiaolangdi reservoir

Fig. 2.5. (iv) Luhun reservoir

Fig. 2.5. The four reservoirs on the mid Yellow River (source: YRCC)

Of the Yellow River drainage basin, 694,000km2 or 92% is located upstream of Xiaolangdi dam site. Thus, Xiaolangdi is located in the key position for flood, sediment control and regulation of the Yellow River.

The Yiluohe river, on the south of the Yellow River's Sanmenxia to Huayuankou section, has two branches of its own. The Guxian reservoir was built on the Luohe river, which is the west branch of the Yiluohe river and the tributary of the Yellow River, while the Luhun reservoir was constructed on the branch of Yihe river of the Liluohe river. Both of the two reservoirs are big according to the standard definition of reservoir magnitude. The construction of the planned dam on the north branch, the Qinhe river, needs to be approved by the state. Thus, it is not part of the present system of the mid-lower Yellow River flood control and management.

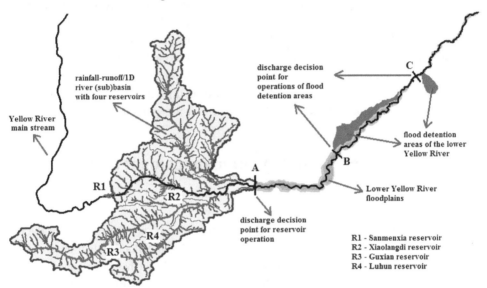

Fig. 2.6. The main flood storage & regulation reservoirs and flood detention areas on the mid-lower Yellow River

The Sanmenxia reservoir has a complex story. It is well known for its severe sedimentation problems (Wang et al., 2005). The key facilities of the dam, including discharge tunnels, were reconstructed due to an alarming rate of loss of reservoir storage capacity and the unacceptable negative impact induced by the rapid upstream extension of sediment deposited in the river's backwater region. The dam was rebuilt to provide more effective release of sediment, and the general operation regime of the reservoir has been changed to achieve a balance between sediment inflow and outflow. The Sanmenxia reservoir is located downstream of the Tongguan hydrologic station, known as the 'entrance' and an important control point of the lower Yellow River flood control and management system. Therefore the floods from the upper section of middle Yellow River will first flow into the Sanmenxia reservoir. Usually, the Sanmenxia reservoir is taken as the upper boundary for the computations for both flood forecasting and flood control management, simply because Sanmenxia reservoir will catch the floods from its upper stream. Thus, Sanmenxia is still regarded as one of the key reservoirs dealing with possible extreme floods of the mid-lower Yellow River, though it has very limited function in regulating small floods. Besides, it also functions as an irrigation and hydropower generation dam, even though some benefits are lower than the original design.

These four reservoirs constitute the main flood control and management regulation system of the middle and lower Yellow River. In case of extreme floods, the Dongpinghu flood detention area located at the right side of the lower Yellow River (the blue area at right bank side in Fig. 2.6) is supposed to be put to use if the regulation by the reservoirs is not enough to control the flow of Aishan hydrologic station to its safe standard. Another flood detention area, Beijingdi (blue area at left bank side in Fig. 2.6) could be used with very extreme floods which are bigger than once-in-a-thousand year floods. Both of these two detention areas are not dealt with in this research for they will only be used with extreme floods after the Xiaolangdi reservoir was put into operation.

2.3 Yellow River flood control management and operational rules

It has been estimated that, for extreme flood events, under the present capacity of the four reservoirs, together with the subsidiary functioning of some flood detention areas, the flood frequency for dike safety of the mid-lower Yellow River has been increased from floods of once-in-60-year to once-in-1000-year approximately (YRCC, 2008). However, for the Yellow River, the safety of the dikes does not mean there is no loss. In fact, there could still be big losses even though there is no dike failure. In the lower Yellow River, there are 1.81 million people living in the floodplains, which are within the dikes (YRCC, 2009). The reason for this is that China has very big population density and people have to make use of every piece of useful land to produce food. At the same time, before the river changed its channel to the present location, the ancient Chinese had already been living in those regions. When the river began to take its course there, the dikes were constructed and people as well as their farmland were enclosed. At present, the flow has become smaller due to human influence in terms of more intensive regulation and more use of water by quick economic development. Smaller flow causes sediment deposits in the main flow course of the channel. Thus, the main course of the river has become higher than the outside and nearby floodplains. This has led to the situation of what is called 'second stage' hanging river (YRCC, 2006a; 2008). So, regulation and control of mid and small magnitude floods are also of great importance for the Yellow River.

In order to deal with the different floods efficiently, with a minimum loss, the principle for controlling and managing the middle and lower Yellow River floods is made according to the present engineering regulative capacity and conditions. Such principle takes the overall interests of the whole mid-lower Yellow River, with all relevant regions and people taken into account, rather than a confined scheme of local people in a particular administrative area. This means that the principle is "to reduce the total loss and risk of mid-lower Yellow River". In this case, the overall interest has the most priority.

This principle is the state-approved and official guideline to conduct the making of operational scenarios/alternatives and the real-time control and management of different floods. It is made through the analysis of historical and design floods and flood frequencies computations, flood magnitude and hydrograph patterns, and a thorough consideration of all aspect involved, namely, flood control and management capacity of reservoirs and detention areas, risk of dike failure, possible flooding of floodplains, possible danger of lives and properties involved, etc.

Generally, for the real-time flood control and management, it is required to follow the strategy and plans, the annual strategy, which have been made before flood season comes according to design and historical floods. These strategy plans will cover the general operational principles and rules. However, as no flood would happen strictly following the historical pattern or the design ones, and all possible related aspects may occur in real-time situations, analyze have to be carried out in real-time operation under real conditions. In that case, it is difficult and unreasonable for a solution to strictly follow the rules. So, an actual action would probably somewhat differ from those in the guidelines. This is also the requirement for the adaptive management approach.

2.3.1 General aspects of flood control management rules on the mid-lower Yellow River

The principle for flood control and management of the mid-lower Yellow River can be conceptually defined on different levels. The top level is the level for the whole middle and lower Yellow River. The flood control principle, at this level, takes into account the interests of the entire region, disregarding the administrative divisions, which include the provinces of Henan, Shandong, Shaanxi and Shanxi. Generally, the implementation of this principle is supervised by the branch organization of the Ministry of Water Resources, i.e. the Yellow River Conservancy Commission. The second level, just next to the top level, is the tributary level, which are rules focusing on a branch of the Yellow River. The second level rules mainly covers the lower level guidelines that the top level does not reach, or does not have to reach. The third level rules are the ones of individual reservoir level, which mainly are the general operational rules made during the design of reservoirs. These designed rules might have been adjusted and modified during later on operation. The priorities of lower level principles are limited to their own valid scope. So the top level rules strictly have the greatest priority.

(1) The three levels of flood control and management rules

Fig. 2.7 shows the three levels of rules, their relationships and control domains. They have different control objectives and different priorities. Generally, the basin level rules will have the highest control priority. This is followed by the branch level ones. The rules for each reservoir have the lowest priorities.

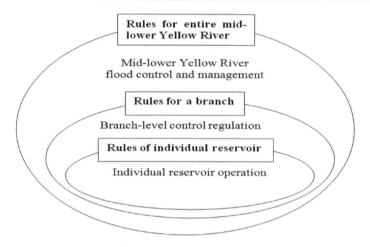

Fig. 2.7. Rules and their control objectives and priorities

In most cases, the lowest level rule will act in the control domain of the upper level ones. However, under extreme conditions, the lower level rules may disregard the higher ones. The typical example is when the reservoir level reaches the highest flood regulation level, the dam safety becomes the highest priority. In this case, the reservoir cannot bear the responsibility for the river or the river branch where it is located. It needs to keep the safety to itself. The high level rules may have some corresponding items to cope with.

(2) Basic terminology for reservoir operation

Different countries and regions will have different reservoir operation rules. Also, the reservoirs will have different operational objectives and targets. So, the terminologies will be different as well. But the general design characteristics are similar, if the reservoirs have similar design operational functions, such as flood control.

Some of the most common items are (See Fig. 2.8):

- Dead pool (storage)
- Dead level
- Food control/regulation pool (storage)
- Conservation pool (storage) (or normal/usable pool (storage))
- Design flood level
- Normal storage level
- Exceptional (check) flood level
- Total reservoir storage

In China, most of the big reservoirs have the function of flood control and management during flood seasons. The two important operational levels are:

- Flood season normal level (FSNL)
- Flood regulation limitation level (FRLL)

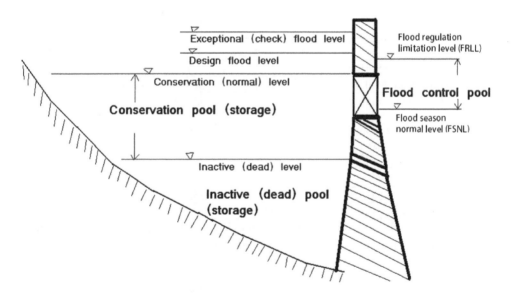

Fig. 2.8. Design/operational characteristics of a reservoir with flood control function

(3) General operational rules for flood control and management

Flood season normal level (FSNL) is an operational level designed to ensure the flood control capacity to be used for regulating the flood when it happens. (Similarly, in the US, there are guide curves) The general rule about FSNL is that the reservoir storage level is not allowed to exceed FSNL during non-flood time in the flood season. During flood regulation, this level becomes the operational target, i.e. the actual operational level should return to the FSNL after the flood regulation operation. At the same time, the water level will rarely go down to below the FSNL due to economic consideration unless with some special situations. An example is that, from the power generation point of view, the level should be kept as high as can be.

When the reservoir begins to regulate a flood, the storage level will rise. The limitation level for this regulation operation is the flood regulation limit level (FRLL). When the storage level reaches the FRLL, the dam safety will be a key issue for the operation. The release is usually increased at above this level. With some reservoirs, the release is required to be in "all open operation" at this condition.

Flood control characteristics of the four reservoirs are listed in Table 2.1 with data of 2009. Due to the heavy sediment load and sedimentation at the bed, the storage capacities of Xiaolangdi and Sanmenxia will change all the time. YRCC issues the latest numbers every year.

Table 2.1 Reservoir characteristics in terms of flood control and management*

Reservoir characteristics (Elevation: MSL)	reservoirs			
	Xiaolangdi	Guxian	Luhun	Sanmenxia
Flood season normal level (FSNL) (m)	225.00	529.30	317.30	305.00
(Designed) flood regulation limitation level (FRLL) (m)	275.00	548.00	323.00	335.00[**]
Present storage of FRLL (billion m³)	10.35	1.02	0.82	5.46[**]
Dam top level(m)	281.00	553.00	333.00	353.00
Catchment size (km²)	694155	5370	3492	688399

Note: * based on data of 2009; ** the Sanmenxia reservoir will only be used to regulate extreme floods due to its severe sedimentation problems. In general, the FSNL will be its operational level during flood regulation process.

Since YRCC defines the operational rules as "internal material", they cannot be given in this thesis. However, necessary explanations will be made about the rules whenever needed.

2.3.2 Flood control computational scheme

The flood control management and decision making process of different river authorities can be different because of management styles and different basins background. The management entails developing alternatives, which may be evaluated by detailed computations. The computation consists of reservoir and river flow routing simulation and inundation prediction. The Yellow River, and also some other rivers in China, have a very big population density in its floodplains, so the flood control and regulation have to be done in a very detailed way. The general processes and main operation are the following (Fig. 2.9):

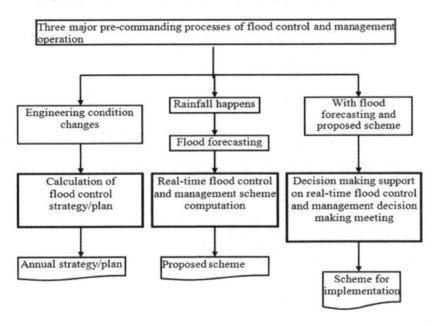

Fig. 2.9. The general computation-related practices of flood control and management

(1) The annual flood control and management plan/strategy

Every year, the conditions of the flood control and management engineering system may have conditions that require changes. For example, a new structure is being built downstream of a big reservoir, and thus the operation principle of the reservoir may need to be changed in order to prevent the construction site to be flooded. This requires modifications of the flood control and management guidelines, to adapt to the new conditions. In order to change the existing flood management strategy, the computation of design floods and/or simulation of historical floods need to be carried out to study the new regulation effects. The computation and the discussion about the computational results cannot always be done in a short time. In general such a computation requires the use of an existing simulation model that will require time to run different solution alternatives. This can lead to different requirements for the developed supporting tools, as was studied and will be emphasized in the present research.

(2) Real-time flood control and management alternative making

During flood seasons, floods of different magnitude would require real-time analysis and computations for every step of the flood development, from the initial forecasting of the flood to the end of the whole flood period. The operation of a reservoir, or a group of jointly-run reservoirs, requires computations such as reservoir routing/regulation, river flow routing/simulation, hazard estimation, etc. Real-time flood control alternatives based on the latest engineering conditions or under certain engineering operational assumptions are made every moment a flood develops. The proposed alternatives need to be analyzed by the stakeholders who play different roles in flood control and management in YRCC's collective decision making meeting.

(3) Adaptive management and decision making

The decision making for flood control and management is crucial to fight floods. Decision making processes involve exchanges of information and computation results, such as reports on meteorological analysis and forecasting, flood forecasting, engineering operation and condition reporting, operational objectives discussions, discussions and analysis of proposed schemes, possible hazard analysis and interactive alternative computation and making of decisions. Out of all these items, the most important is the elaboration of an alternative for the real-time operation of the reservoirs. At such meetings, excellent and robust software is needed, in order to carry out quick response evaluations of any requirement by the decision makers.

In addition to the above three practices, the analysis of design flood in some planning departments also lead to the requirement of similar computation system. However, if the computation is done considering just the design flood, it will not be in real-time manner. In case that design floods are commutated by a DSS there is no need of immediate response to the user operations, however such systems should be able to work not only in case of project design and planning but also for real-time flood control and management operations. This would imply a different approach in the architecture of such a system.

A simplified procedure for discussion and decision making meetings in China's large river authorities is shown in Fig. 2.10. The meeting is held by decision makers, experts, and relevant people in charge of flood forecasting, hydrologic information, engineering management, etc. In this meeting, all latest issues that are key to the flood situation, will be studied and examined before decisions are made. The processing in flood control and management is not a pure constraint problem. Usually, at any moment, the decision makers are very clear about the rules and goals that need to be implemented in crisis situations. Real-time flood control and management implies that a decision meeting could be held at any moment as the flood develops, so a series of meeting different flood development stages could be held. A solution worked out automatically by the computer can only be a reference for the decision makers or meetings. From adaptive management point of view, no computer can easily work out an ideal solution. Therefore, the mechanisms to support the decision making process are the most important function of the flood control and management software system (Li, 2008).

Generally, during a decision meeting, the main topics develop as the discussion takes place and the issues are presented. The main goal is to reduce the flood impact. Then, more specific problems are addressed. In this process, the DSS should be limited to those that are directly needed rather than on all processing functions of the system. Just like the concept of 'constraint programming', as the whole process goes on, a lot of inessential factors are to be discarded. At last, the main decisive factors are studied with the help of the software system, but the final decision is made collectively by the decision makers and the experts present in the meeting.

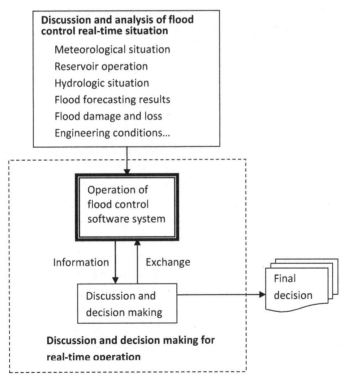

Fig. 2.10. The general procedure of discussion and decision making during the meetings of China's large river authorities

2.4 International developments in flood management

The way human beings deal with floods has been in continuous progress and innovation throughout time. Engineering systems have been built to control and manage floods. Non-structural measures are used to provide early warning. More and more, new strategies and measures are made to optimize flood management. In the past, flood control was the general way for people to tackle floods. But now, people consider floods not only as a threat, but also as a special resource which could be useful. Therefore, the concept from flood control to flood management now starts to have influence on the activities of river authorities dealing with floods. The ways people tackle floods have now a much greater variety than ever before. Also the measures to deal with and manage floods, become more and more delicate and refined. Thus, better tools are required for better management and decision making.

2.4.1 The International Flood Initiative

UNESCO and WMO are aware of the significant achievements that have been made in flood management in recent years and also of existing opportunities to elaborate practical solutions. The United Nations University (UNU), the International Association of Hydrological Sciences (IAHS) and the International Strategy for Disaster Reduction (ISDR) agreed to launch the International Flood Initiative (IFI). At the same time, other UN agencies dealing with various facets of flood management were invited to join the Initiative (WMO-UNESCO, UNU and IAHS, 2007).

IFI was set up to develop and share knowledge on all flood related activities which include monitoring, network design, statistical analysis of floods, real-time forecasting and flood modeling. At the same time the initiative focuses on assessing community vulnerabilities and risks, as well as on their associated factors, including poverty reduction, population growth, migration to urban centers and mega-city development. Interdisciplinarity is the core philosophy of all actions and activities undertaken by IFI. All scientific knowledge including economic and social sciences, as well as new technologies, particularly remote sensing and ICT, are harnessed in every step of the Initiative. Emphases is put on involvement of all stakeholders in flood management. Various development activities within a basin that affect and are affected by floods are carried out under different sectors and administrative jurisdictions. Interdisciplinary and intersectoral coordination and understanding is essential for designing and implementing institutional reforms and participatory stakeholder processes that promote fair and effective flood management policies (IFI website, 2011).

Based in Japan, the International Center for Hazard and Risk Management (ICHARM) also focuses on integrated water resources management (IWRM). According to

ICHARM, integrated water resources management is a step by step process of managing water resources in a harmonious and environmentally sustainable way by gradually uniting stakeholders and involving them in planning and decision making processes, while accounting for evolving growth, rising demand for environmental conservation, changes in perspectives of the cultural and economic value of water, and climate change.

'Integrated River Basin Management' is referred to in the context of implementing IWRM for the provision of water services at the river basin level. It was defined by the Global Water Partnership (GWP) as 'a process which promotes the coordinated development and the management of water, land and related resources, in order to maximize the resultant economic and social welfare in an equitable manner without compromising the sustainability of vital ecosystems'. The river basin approach will focus on implementing IWRM principles on the basis of better coordination among operation and water management entities within a river basin, with a focus on allocating and delivering reliable water-dependent services in an equitable manner.

The tri-annually held International Conference on Flood Management (ICFM) is the recurring international conference entirely focused on flood related issues. It is designed to bring together practitioners and researchers alike, including engineers, planners, health specialists, disaster managers, decision makers, and policy makers engaged in various aspects of floodplain management. It provides a unique opportunity for these various specialists to come together to exchange ideas and experiences. Therefore, the evolution of concepts and activities dealing with floods (ICFM5 website, 2012).

The first time it was called International Symposium on Flood Defense held in Kassel (Germany, 2000). The later conferences were held in Beijing (China, 2002), Nijmegen (The Netherlands, 2005), Toronto (Canada, 2008), Tokyo (Japan, 2011) all emphasizing flood defense measures with each successive event, evolving towards more integrative approaches, including risk, vulnerability and capacity development. The 5th International Conference on Flood Management (ICFM5), held in Tokyo, Japan in 2011, marked the continued advancement of flood management practices and policies around the world. At this conference, the name was changed from 'control' as used in the previous four events to 'management', which reflects the more integrative approaches to flood management that nations are increasingly employing. The ICFM5 theme was 'Floods: From Risk to Opportunity', which reflects the continued trend towards a broader understanding of how we collectively make use of the opportunities provided by floods and flooding, cope with risks and plan for response to flood events.

The themes of all five events reflect how the concepts of flood control and management have been changing over the past decade or so.

2.4.2 The EU Flood Directive

The Directive was proposed by the European Commission in 2006, and was published in the Official Journal on 6 November 2007. The aim of this directive is to reduce and manage the risks that floods pose to human health, the environment, cultural heritage and economic activity. The Directive required EU member states to first carry out flood risk analysis, the preliminary assessment of risk of floods by 2011 to identify the river basins and associated coastal areas. Then the flood risk management maps can be made by the year 2013. Also management plans should be made which focus on prevention, protection and preparedness by 2015. The directive applies to both inland waters and all coastal waters across the whole territory of the EU. The directive is carried out in coordination with the EU Water Framework Directive, notably by coordinating flood risk management plans and river basin management plans, and through coordination of the public participation procedures in the preparation of these plans. The directive requires that all assessments, maps and plans prepared shall be made available to the public.

The EU member states coordinate their flood risk management practices in shared river basins, including third counties, and agreed in solidarity not to undertake measures that would increase the flood risk in neighbouring countries. These states also take into consideration their long term developments, including effects of climate change, as well as sustainable land use practices in the flood risk management cycle (Official Journal of the European Union, 2007).

2.4.3 Flood Control 2015 Program in The Netherlands

The Netherlands is a country that has a very long history in fighting floods. The modern history of this country shows great efforts and success in dealing with sea storm surges and river floods. Dutch companies and knowledge institutes have taken compressive measures to optimize flood risk management. The Flood Control 2015 program combines water management and information management in an innovative way. Facing the threats of rising sea level, sinking coastal land, and extreme weather conditions, The Netherlands aims to properly anticipate these problems, which is why the Dutch government launched the Flood Control 2015 initiative.

Forecasting systems developed within the Flood Control 2015 programme are to ensure that better information reaches the right place more quickly. Current flood protection focuses more on strong dikes, but the greatest gain lies in making the total system smarter: the dike, the decision-maker, and their environment. Flood Control 2015 integrates them into more advanced forecasting and decision support systems. Then, flood protection becomes more transparent, quicker and better. And the cost of managing water systems could be reduced at the same time (Flood Control 2015 website, 2012). In the Flood Control 2015 program, the construction of a flood control room is one of the key tasks, directly for decision making and command. Once decisions are taken, it is extremely important that there is real-time insight into the various scenario's. The room, like the collective decision meeting room in YRCC, will constantly monitor water levels and levees. At this time, in the Flood Control Room, new techniques can be tested and disaster scenarios can be simulated. According to the programme plan, this room will also be used to train water management professionals.

The Dutch Delta Program is a nationwide program which enables the national government, provinces, municipalities and water boards to work together with input from social organizations and the business community. The main objective is to protect the Netherlands from flooding and ensure adequate supplies of freshwater for future generations. This programme consits of nine sub-programmes: Safety, Freshwater, New Urban Developments and Restructuring, IJsselmeer region, Rhine Estuary-Drechtsteden, Southwest Delta, Rivers, Coast and Wadden region. This programme is to plan protection measures for the Netherlands against flooding and also to ensure adequate supplies of fresh water, now and in the future. Making this plan involves many issues and factors both around spatial, urban planning, the economy – and on the level of agriculture, nature and recreation (Delta Program website, 2012).

2.4.4 Flood Risk Management Consortium in the UK

In the UK, the Flood Risk Management Program was initiated to provide a key focus for many national research programs on flooding from rivers, estuaries and the sea that pose a serious threat to many people and property. The Flood Risk Management Research Consortium (FRMRC) is an interdisciplinary research Consortium focusing on some recently identified strategic research issues which investigate the prediction and management of flood risk and is the primary UK academic response to this challenge.

The Flood Risk Management Research Consortium covered two phases. The first was launched in February 2004 and the 2nd phase ended in 2011. The consortium combines the strengths of fundamental and near-market research in a truly multi-disciplinary program. This program was able to enhance the understanding of flooding and improve the ability to reduce flood risk through the development of sustainable flood management strategies. Implementation of these strategies requires a flexible, practical and robust decision support system for real-time operation (Cluckie, 2012).

2.5 Software systems for operational management

In YRCC, software development for river management started to develop in early 1980s. These developments cover applications in flood forecasting, flood control management and decision support, water quality simulation, water allocation and management, GIS-based engineering monitoring and management, water and soil erosion monitoring and simulation, 2D flood simulation of floodplains and flood detention areas, etc. Flood forecasting system and flood control decision support system were among the pioneering systems and entail a lot of developing efforts. More broadly, a software environment for different sectors of Yellow River management was suggested by Mynett and de Vriend (2005). This environment was proposed to have four layers: (i) the support layer, (ii) the functionality layer, (iii) the application layer and (iv) the usage layer (Mynett and De Vriend, 2005), covering simulations of flow, water quality, estuary, morphology, etc. The Digital Yellow River project is such a platform incorporating different packages for various river managements.

Flood forecast and management are the most popular tools dealing with floods. Unlike the flood forecasting software tools, which are helpful in real-time operation, the flood control and management software systems are still not yet satisfactory at various river authorities (YRCC, 2006b; ICMWR, 2006; CWRC, 2006). Hence, in many river and reservoir authorities in China, new systems are still being planned or developed, in order to have more practical and successful applications.

The difficulty in developing a good software system for flood control and management comes from its specific functionality (Jonoski and Popescu, 2011; Popescu et al, 2012). On the one hand, as more and more new flood control and management concepts and practices are developed, adaptive management becomes more and more sophisticated. On the other hand, adaptive management and decision support demand a software tool, which is very different from a normal simulation or modeling tool, which most of the application software developers and researchers are used to. However, many developers and researchers simply take a normal simulation or modeling system as a management and decision system. They are not aware of the fact that an ordinary simulation system or modeling tool is not a management system, nor a decision system. For general research purpose, perhaps a simulation system is eligible in terms of functions and behaviors. However, adaptive management and decision support need better robust functions, better system performance, good interaction mechanism, better user interfaces.

As for comparison, a reliable flood forecast software system with embedded good forecasting models would be quite adequate for real-time flood forecasting which can be regarded as a simulation problem, while for flood control and management software, the precise calculation of mass balance in reservoirs and accurate routing in river channels does not necessarily imply the software meets the real-time operational requirement. The interaction between man and machine, the mechanism supporting the decision making process, the output style and contents, and system performance, are all crucial for such a system.

The uniqueness and special requirements come also from the users' character. Decision makers, experts, as well as decision making meetings, are different from ordinary users doing simulations and modeling. These users will not be interested in low level data access and presentations. Decision makers need an efficient support for discussions and analysis to meet the "what-if" style information exchange. Low-level data, or information, should not be presented in these meetings.

Since most of the users are experts in flood management and have enough knowledge on the conditions and rules, the user interfaces should be made rather friendly and "hydro-tradition-oriented". Interfaces nowadays available as common software component tools may not be suitable for these kinds of users. The computation result should be presented at high-level abstraction and include graphics. The system should have quick or even immediate response to any reasonable change of the engineering control parameters. A discussion, or a decision making meeting, may lasts for half hour to few hours. In this circumstance, if a relevant simulation runs dozens of minutes or more would be unacceptable, because it stops the on-going discussions or decision making process.

At the Yellow River Conservancy Commission, the collective flood control and management meeting could be held at any time when a flood develops. Usually this meeting takes one or few hours. There will be many exchanges of ideas among decision makers, experts and relevant department heads. The ideal response time for a good software should be within seconds. A few minutes may be acceptable to wait for a result for a "what-if" alternative development process. If the meeting takes one hour, then hours of runtime of software would be of no value to the meeting. No experts or decision makers would wait for hours at the flood control room or decision meeting for a slow response.

It must be mentioned here that the users may not operate the system themselves. For example, the decision makers may not operate the system physically by themselves, but they should be regarded as "direct" users of the system.

The Yellow River Conservancy Commission started the development of flood control applications in mid 1980s. Supported by the state and National Flood Control and Drought Relief Headquarters, YRCC developed the first formal flood control software system in 1995, named "Yellow River Flood and Ice Flood Control Decision Support System" (Cui et al, 1998). The system was planned as a decision support tool rather than a simulation system. The system was developed to support the real-time operation and decision support of Yellow River floods in summer and ice flood in winter. The main functions were rule-based alternative making, decision support in discussions and collective decision meetings. This system was based on the DOS system and was abandoned when the Microsoft Windows became popular.

Later, a similar system was developed during a Sino-Finland cooperation project on a UNIX platform. The UNIX application was abandoned not long after it was put to use for the simple reason that Microsoft Windows platform was and still is much more popular.

The present system, the Yellow River Flood Control Regulation System was developed by Hohai University, China. It was a part of the National Flood Control and Drought Relief Commanding System (NFCDRCS). The NFCDRCS, which focuses mainly on auto gauging, flood forecasting and flood management, as organized by the National Flood Control and Drought Relief Headquarters and funded by the state. It is up till now the biggest system of its kind in China, and also a national program covering all the important gauging points and areas of the entire country. The NFCDRCS is aimed at using ICT to upgrade the flood gauging, data collecting and processing, flood forecasting, flood control and management hardware and software systems in China's main rivers and lakes systems, especially for the seven big basins, i.e. the Changjiang (Yangtse) River, the Yellow River, the Zhujiang (Pearl) River, the Songliao River, the Haihe River, the Taihu Lake and the Huaihe River basins.

Though many improvements can be seen when compared with the previous old versions, there are still some common drawbacks, which are typical, not only in YRCC, but also in other large river authorities.

2.6 The need to improve decision support tools

In this section, the DSS software presently used in YRCC is examined and conclusions are made regarding the shortcomings and problems of using it. Then need for improvement is analyzed. This will be an important basis for developing better tools.

2.6.1 Adaptive management

Adaptive management as a concept for simultaneously managing and continuous learning has been around for several decades (Holling 1978; Walters 1986). Adaptive management is a systematic process for continually improving management policies and practices by learning from the outcomes of previous situations (Williams, 2011). The foundations of adaptive management rest in many fields, but its initial presentation as a natural resources management paradigm was proposed in the 1970s, when it was offered as a way to help managers take action when faced with uncertainties. The method helps to elaborate management strategies capable of responding to unanticipated events (National Research Council, 2004). Holling (1978) used this term in his book reporting on the findings from a workshop at the International Institute for Applied Systems Analysis in Vienna, Austria. Walters and Hilborn (1978) also used this term in a paper named 'Ecological Optimization and Adaptive Management', which explored the consequences of uncertainty by examining various optimization analyses for managed populations, beginning with deterministic optimal control models that presume full knowledge and ending with adaptive control models that presume almost complete ignorance. They concluded that actual practice always involves a richer set of objectives, constraints, and hedging activities than any model can do, since model are simplifications. They recommend to break out the passive mode and learn to treat the acquisition of functional information and indeed the whole management process as fundamentally experimental activities requiring active planning and judgment.

In addition to natural resources management, applications can also be found in other fields, including systems theory (Ashworth, 1982), business (Senge, 1990). Therefore, its origin is rooted in parallel concepts from a variety of perspectives, but in natural resources the term simply means 'learning by doing' or 'adapting based on what's learned' (Walters and Holling, 1990). 'Active' adaptive management employs management programs that are designed to experimentally compare selected policies or practices, by evaluating alternative hypotheses about the system being managed. The key characteristics of adaptive management include: (i) acknowledgement of uncertainty about what policy or practice is "best" for the particular management issue; (ii) thoughtful selection of the policies or practices to be applied; (iii) careful implementation of a plan of action designed to reveal the critical knowledge; (iv) monitoring of key response indicators; (v) analysis of the outcome in consideration of the original objectives; and (vi) incorporation of the results into future decisions (Nyberg and Taylor,1995; Nyberg, 1998).

According to the National Research Council (2004), adaptive management promotes flexible decision making that can be adjusted in the face of uncertainties as outcomes from management actions become better understood. Careful monitoring of these outcomes can advance scientific understanding and helps adjust policies or operations as part of an iterative learning process. It is not a 'trial and error' process, but rather 'learning while doing'. Adaptive management does not represent an end in itself, but rather a means to achieve more effective decisions and enhanced benefits. Gray (2005) describes the so-called behaviour-based adaptive management and states that it can be envisioned as a cyclical process of learning by doing using any number of feedback mechanisms available. The cycle can be sophisticated or simple, subject to the approach selected. Behaviour-based adaptive management is perhaps best understood as a continuum of learning tools ranging from: (i) reactive, event-by-event, trial-and-error decision-making;

to (ii) single policy design, implementation, monitoring, and modification as required (passive); to, (iii) multiple policy evaluation using sophisticated active experiments and comparative analyses (Hilborn, 1992). Rist, Campbell and Frost (2012) listed eight key components: (i) participation of those outside the management institution in order to manage conflict and increase the pool of contributions to potential management solutions; (ii) defining and bounding of the management problem, including the setting of management objectives; (iii) representing existing understanding through system models that include assumptions and predictions as a basis for further learning; (iv) identifying uncertainty and alternative hypotheses based on experience; (v) implementation of actions/policies to allow continued resource management or production while learning (reducing uncertainty); (vi) monitoring the effect of implementing new policies; (vii) reflecting on, and learning from, monitoring results, comparison with original expectations in order to revise models and/or management actions; and (viii) iterative repetition of this cycle (points i–vi above) so that uncertainties are reduced and management outcomes are improved over time.

Applications have been shown extensively in literature. The U.S. Department of the Interior recently listed a focus on practical applications in the areas of importance to its managers: climate change, water, energy, and human impacts on the landscape (Williams and Brown, 2012). It presents adaptive management as a form of structured decision making, with an emphasis on the value of reducing uncertainty over time in order to improve resources management. The first half of the guide covers the foundations and challenges of adaptive management, and the second half documents examples that illustrate the components of adaptive management. Application areas of adaptive management include:

(i)   decision support for resources management (Westphal et al, 2003); natural resources management (Stankey et al, 2005; Allan, 2007; Williams, 2011; Williams and Brown, 2012); water resources project planning (National Research Council, 2004); water resources management (Custer and Sojda, 2006; Mysiak et al, 2010; Engle et al, 2011); management of connected groundwater-surface water resources (Brodie et al., 2007);

(ii)   fishery management and fish passage design (Weber, 2007); management of threatened and endangered species (Runge, 2011);

(iii)   forest management (Millar et al, 2007);

(iv)   adaptation to climate change (Dessai and Hulme, 2007); management of climate change problems (Gheorghe, 2008; Milly et al, 2008);

(v)   environmental assessment and management (Holling, 1978);

(vi)   management of ecology and ecosystems (Walters and Holling, 1990).

It can be concluded that, though there are still some arguments about the concepts and the actual implementations, adaptive management has become one of the basic approaches for managing natural resources and the environment.

For flood control and management, the rules are conventionally made based on historical information and previously established design values. However, future flood events may differ due to e.g. effects of climate change or real-time meteorological or hydrological conditions. Uncertainty exists in all components and floods may vary in extent in terms of peak discharge, arrival time, etc. Therefore, control and management decisions have to be made according to actual and real-time flood situation. From practice on the Yellow River, it can be said that the rules are general guidelines for operational control and management rather than universal solutions for all situations. At the same time, simulation models and analysis tools may have errors in extreme events that are essentially different from historical design. This also brings uncertainty into the system. The above considerations imply to - at least - consider an adaptive management approach.

The following subchapter addresses the implications for the software architecture of such a system which facilitates an adaptive management approach.

2.6.2 Adaptive decision support software architectures

Adaptive management brings challenges to DSS tools. Looking at the software tool being used by YRCC and by other river authorities as well, it is found that most of them run in the common way that a simulation model does: (i) determine and prepare the input in the form of boundary and initial conditions, (ii) run the software, and lastly, (iii) check the results. This may be appropriate for scientific research where the system runs in a procedural way and time need not be a limiting factor. However, in real-time operation for flood control and management, there are cases where the system needs to perform very fast to support decision making meetings. Studies of various existing DSS systems in China trigger the following conclusions:

- decision support systems require different ways of interaction and faster response times compared with general model-based simulation systems used in research;

- developers of such systems should have a clear idea about this difference;

- developers should have a good knowledge and understanding of the actual process of real-time operational management and decision making;

In YRCC, staff in charge of system operation are generally reluctant to run their system directly in front of the decision makers or during decision making meetings. The preferred way is that they obtain their results (proposed alternatives or scenarios) back in their own departments, and then show the results in hard copies to the decision makers or during the decision meetings. In real-time operation, especially when time (e.g. lead time provided by forecasting) is very limited, this is not efficient.

2.6.3 Internal execution pattern and operational mechanism

As an example, after review of its system in operation, the Changjiang (Yangtze River) Water Resource Commission (2006) concluded on the use of the decision support system for Yangtze River flood control and management: *"......a big gap still exists between actual operational requirement and the available functions of the existing system, especially in the system operation mechanisms and man-machine interactive decision support; knowledge and experience need to be improved to deal with structured and semi-structured decision problems to develop really useful and advanced DSS."*

Hence, it is necessary to study the present situation and improve the system's practicability.

2.6.4 Flexibility on function compositions and rule management

River management authorities have been in a continuous effort to seek and practice new ideas and principles to be applied for river management. An example of such effort is the change from flood control to flood management or even from flood control to integrated water resource management (e.g. Kahan et al, 2006). In the past, the Yellow River Conservancy Commission had continuously introduced new ideas and guidelines for the management of the river. Also, as a result of fast development in the engineering system, changes of engineering conditions happen more and more quickly on the river.

Thus, the principles and rules of flood control and management are supposed to slightly change each year. These changes lead to the need of modifying the functions of the system and also require development of new models. Hence, the developed DSS should be designed in such a way that it would be easy to add a new function to meet a new user requirement. On the Yellow River, the annual strategy will be issued every year and big or small changes are likely to be made to the old one so that the strategy is more purposeful with respect to the new management and engineering conditions.

The present DSS system was coded with rules which cannot be changed easily without developers editing the source codes and recompiling it. Both the architecture of the system and the low-level computation blocks were programmed according to the present conditions and could not be changed easily. The users will find inconveniences or difficulties whenever they need to make some changes to the computation targets or processes. Adjusting or changing the operational rules also demands a flexible system structure for the composition of functions.

Generally, the rules control the computation procedures no matter what the control objective processes are: a single reservoir, a river branch system with reservoirs, or a basin-level flood control and management. If the system deals with only one single reservoir, the rules will be the ones that belong to the reservoir. If one or more reservoirs together with the relevant river reaches form a flow regulation system, then the rules will be at a higher level. Thus, the rules will control the computation for the reaches or even the entire basin.

For real-time operation, friendly software tools for applying new rules are to be developed. Both the HEC, DHI and Delft simulation software tools emphasize this ability. The so-called 'user-scripted' rule editing and 'Visual Basic macros' are being used to provide the user with the function of adjusting or changing the operational rules.

2.6.5 Evaluation of alternatives for flooding simulation

The suspended nature of the lower Yellow River has made the flood control and management extremely crucial for people living and farming in the floodplains. Due to the high population density on these floodplains, which are the residential place for nearly two million people, the pressure on the environment and flood protection is high. The flow peaks and hydrograph shapes are, in the current practice, the key factors to determine whether these floodplains are safe or not.

It is critical to regulate the flood forming in the mid Yellow River that is propagated to the lower reach. Smaller peak discharges, smaller high-level volumes or optimally regulated hydrographs might be able to reduce the inundation extent in the lower Yellow River floodplains. Analysis and decision making of flood control and management of floods involve evaluation of multiple options and criteria related to least flood volume, control peak discharges at certain points, and least inundation areas. All these demand a flood simulation model in lower Yellow River for determining inundation depths.

The YRCC has been developing a large 2D simulation tool for lower Yellow River. The project is called YRCC Model project which aims at simulating flow and sediment transport process in the lower Yellow River (Zhu et al, 2005). Because of the size of the lower Yellow River, which has a length of around 800 km and a width from several kilometers to 20 km, this model has always been a big application which needs an extraordinary computational power.

The complex nature of flow and sediment content also contributes to the long running time of the model. Now in YRCC, a super-computing center has been set up, for the applications like the YRCC Model. Still, a general run of the model would take dozens of hours or even days. This makes it hard to be used in the flood control discussions and decision making meetings which, in the analysis of flood, require an immediate response based on the 'what-if' analysis. Therefore, 2D or 1D2D coupling simulation models should be explored for the prediction of flooding and evaluation of reservoir operation alternatives for the upper part of the lower Yellow River.

2.6.6 Other requirements for improvement

(1) User interfaces in real-time operation

The user interface is the 'window' of a software system to serve the users. User interfaces can determine whether a system is actually being used. Bad user interfaces can easily spoil the entire system. The present system at YRCC is implemented based on the typical Windows interface style, and there is no consideration for the special features required for flood control and management, nor for relevant decision making procedures. Graphical GIS-based user interfaces are required to support an adaptive management approach.

It is clear that the computation and decision making process of flood control and management has its own and special characteristics; the software's operational style should be designed accordingly.

(2) Graphic analysis tools

Decision makers are not used to do modeling, they are the ones to whom modellers should show their results and findings, so that a good decision can be taken. They should not be given low-level or raw data. Graphical presentations can be very helpful to them. With a confined lead time from forecasting, the decision making process has to be as efficient as possible. In addition to the general information that is delivered to the decision makers, the information that is produced interactively in real-time discussion is also very important. There should be comprehensive analysis tools available, based on graphical operating modes.

2.7 Summary

Reservoirs, together with river channels, constitute a complex network system for storing, regulating and conveying flows. In such systems, the operation of reservoirs is essential for the control and management of floods and the protection of people and property in relevant regions. The management discussions and collective meetings of decision makers and experts responsible for making operational scenarios, are of utmost importance as a flood develops.

The general process is that the latest information on hydrological forecasts and reservoir conditions and other relevant issues are presented and discussed in a high-level meeting. After the overall goals are set, the more specific problems are then focused on, such as which reservoirs should be operated and how to regulate the flood. The less important issues will be agreed upon by technical experts, but the more crucial decisions are discussed and agreed collectively in the decision making meeting.

Real-time flood control and management also implies that a decision meeting could be held at any moment as the flood develops. A complete flood process could involve a number of meetings in different flood development stages. A solution worked out automatically by the computer can only be a reference for the decision makers or meetings. Decisions have to be made based on analysis of real-time situations.

From an adaptive management point of view, no computer can easily work out an ideal solution. Therefore, the mechanisms to support the decision making process are the most important functions of a flood control and management software system. After many years of development of supporting tools, decision makers and end-users of such systems are still not satisfied. There is an obvious need to develop new tools which should be able to achieve more efficient support for the decision making process, which usually takes place within only a few hours of precious lead time.

3 Multi-reservoir-based flood regulation

This chapter presents a simple example of multi-reservoir-based flood control regulation. The basic aspects related to reservoir regulation and reservoir routing processes are presented. The evaluation of operational alternatives is illustrated with worked examples. Taking into account the requirements of a DSS tool and the adaptive management approach, the basic computation and routing methods are analyzed and discussed.

3.1 Basic reservoir regulation equations

Reservoir routing, or reservoir storage routing, is the basic process of calculating the passage of a runoff hydrograph through a reservoir. To be more specific, the routing entails the computation of the resultant outflow hydrograph (q), storage (V), as well as the corresponding reservoir water level (h), during the passage of an inflow hydrograph (Q) through a reservoir.

The general simple relation among inflow, outflow and storage, is: over a given period of time, inflow minus outflow is equal to the change of storage (Chow, 1964). This relation is the mass conservation equation of the reservoir's in-pool water storage. Based on the above stated principle of conservation of mass, at any moment, the reservoir storage (in-pool water volume) equation can be written as:

$$\frac{dV}{dt} = Q(t) - q(t) \qquad (3.1)$$

Where Q(t) is the inflow into the reservoir, as a function of time; q(t) is the outflow (release through all outlet facilities in operation) as a function of time, V is the reservoir storage (in-pool water) volume; and t is time. In general, computation is carried out, by transforming the above equation in a finite difference one, and solution is applied in small time steps, of length Δt, for a given time series of input discharge:

$$V_{t+1} - V_t = \left(\frac{Q_t + Q_{t+1}}{2}\right)\Delta t - \left(\frac{q_t + q_{t+1}}{2}\right)\Delta t \qquad (3.2)$$

where Q_t, Q_{t+1}, q_t, q_{t+1}, V_t, V_{t+1} are instantaneous values of inflow, outflow discharges and storage for time step t at the start, and t+1 at the end moment of computation, over the interval Δt.

A close analysis of equation (3.2) reveals that it cannot be solved without adding to it a second relation, because it has two unknowns, V and q. Many methods have been developed in time, depending on the second equation that is taken into consideration and the method of solution. One of the most known developed routing methods based on equation (3.2) was the one developed by Puls of United Stated Army Corps of Engineer (Puls, 1928). Later on, the method was modified and sometimes called Modified Puls Method, which has the following form:

$$Q_t + Q_{t+1} + \left(\frac{2V_t}{\Delta t} - q_t\right) = \left(\frac{2V_{t+1}}{\Delta t} + q_{t+1}\right) \tag{3.3}$$

In equation (3.3), the unknowns in the routing calculation, Vt+1 and qt+1, are all on the right side of the equation, while all known values are on the left side of the equal sign. Thus, another relation between V and q can be used in accordance with the terms $2V/\Delta t - q$ and $2V/\Delta t + q$, namely the relations in the forms: $(2V/\Delta t - q) \sim q$ and $(2V/\Delta t + q) \sim q$.

Similar to equation (3.3), another very popular form is:

$$\frac{Q_t + Q_{t+1}}{2} - q_t + \left(\frac{V_t}{\Delta t} + \frac{q_t}{2}\right) = \left(\frac{V_{t+1}}{\Delta t} + \frac{q_{t+1}}{2}\right) \tag{3.4}$$

With equation (3.4), only one V and q relation is used. This is the $(V/\Delta t + q/2) \sim q$.

Equation (3.4) is the storage-release relation (V-q relation). Another relation, storage level H and storage volume release V always has a one to one correspondence, in which case the H-q relation can be used for routing instead of V-q.

The first attempt to solve the governing equation (3.1) analytically was done by Fenton (1992) who starts by writing equation (3.1) in the even more general form

$$\frac{dV}{dt} = Q(t) - q(t, h) \tag{3.5}$$

In equation (3.5), the release q is defined as a function of both time (t) and water level (h) in order to be able to also consider the controlled release by given value and release by opened gates or spillways. Fenton then uses the relation between elevation h and in-pool water surface area A to obtain an analytical solution.

In a special test case, he obtained the solution as follows

$$q = Q_o + e^{-ct}\left[q_0 - Q_0 + \frac{s!\,cP}{(f-c)^{s+1}}\right]$$
$$-\frac{s!\,cP}{(f-c)^{s+1}}e^{-ft}\left[1 + (f-c)t + \cdots + \frac{[(f-c)t]^s}{s!}\right]$$

where Q_0 and q_0 are initial values of inflow and outflow at time t=0; s represents the storage as an integer value. If it is not an integer, a solution in terms of incomplete gamma functions can be found. In the above analytical solution several parameters need to be calculated for each particular reservoir taken into consideration.

However routing a reservoir may be solved differently for long term planning and design of water systems. For such systems Loucks and Van Beek (2005) use the mass-balance equation over a time period T that has one extra term L_T:

$$V_T + Q_T - q_T - L_T = V_{T+1} \tag{3.6}$$

where L_T represents the evaporation and seepage losses. This term should be considered in especially mid-long term planning and management of water resources. Based on this equation, sophisticated strategies and methods have been developped and suggested by Loucks and Van Beek (2005). Depending on the planning and management goals, the objectives and constraints can be formulated and, based on that, relations can be built for reservoir operational yield. Together with equation (3.6), these provide the projected output in time.

Wurbs (2012) presented a comparative review of river/reservoir system modeling capabilities that integrates the Texas experience with the US nationwide endeavors to develop and implement generalized models. The objective is to assist practitioners in selecting and applying models in various types of river/reservoir system management situations and to support research in continuing to improve and expand modeling capabilities. This paper reviewed much literature on reservoir/river system models and then focuses on generalized modeling systems that have been extensively applied by water management agencies in a broad spectrum of decision-support situations in Texas and elsewhere. Several modeling systems are suggested as being representative of the state-of-the-art from a practical applications perspective. Modeling capabilities are explored from the perspectives of types of applications, computational methods, model development environments, auxiliary analyses, and institutional support (Wurbs, 2012).

3.2 Yellow River multi-reservoir flood regulation

From the computation point of view, the multi-reservoir-based flood control and management software system, which is similar to an integrated river basin management system, consists of reservoir routing, river flow simulation and rule-based flow control (Li, 2008). The reservoir regulation is the mass balance calculation of the in-pool water storage, however, a fine simulation of the reservoir pool, for management and decision will include many elaborate calculations, such as keeping a fixed storage level, release of a constant discharge, release of a given time series of discharges, etc (Hassaballah et al, 2012). The organization of these components and the management of the controlling rules dominate the multi-reservoir regulation calculation process. Fig. 3.1 shows the logical relationship among the rules and reservoir regulation computation for the multi-reservoir river system of mid-lower Yellow River. The basic computational routines are based on routing methods of various operations. Logically the routing methods are at the bottom of the entire computation. The rules of different control levels are the guidelines to link different basic computations and also control the computational flow.

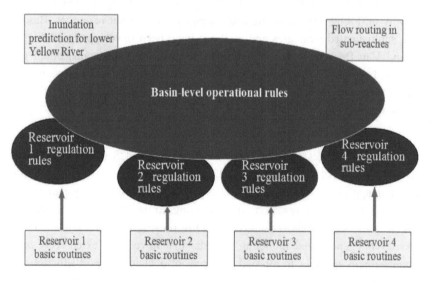

Fig. 3.1. Logical relationship among the rules and regulation computation

Fig. 3.2 shows a sketch of the four Yellow River reservoirs and their control target point and the Huayuankou hydrological station, which is located approximately at the upper boundary of the wide lower Yellow River channel. From the overall flood protection interest point of view, the Huayuankou's hydrograph is the resultant hydrograph from the operation of the four reservoirs. At the same time, this hydrograph is the upper boundary inflow for the lower Yellow River. In fact, the general basin control rules are largely based on the forecasted hydrograph of the Huayuankou hydrological station.

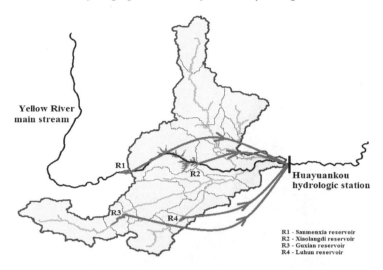

Fig. 3.2. The four reservoirs and the downstream flow control target point

Fig. 3.3. shows an example of a simplified hydrograph at Huayuankou station, obtained by overlapping the four sources, namely, the rainfall runoff forecast (in red color) for Xiaolangdi to Huayuakou region, the flow (in purple color) from Heishiguan of Yiluo river, the flow (in blue color) from Qinhe river and the flow (in red color) from Xiaolangdi's release. As an example, it can be seen that Xiaolangdi captures its inflow and does not release during time when its release will contribute to the Huayuankou's peak discharge. It releases its flood volume storage to the tail section of the Huayuankou hydrograph. This makes the lower Yellow River facing a less high flow time and therefore it is safer.

The hydrograph of Huayuankou is a direct control target of the multi-reservoir regulation. By experiences, the decision makers and experts will have a general guess about the possible flooding in lower Yellow River. In YRCC, some inundation statistics have been made after historical events to make a kind of experience-based inundation maps for later use.

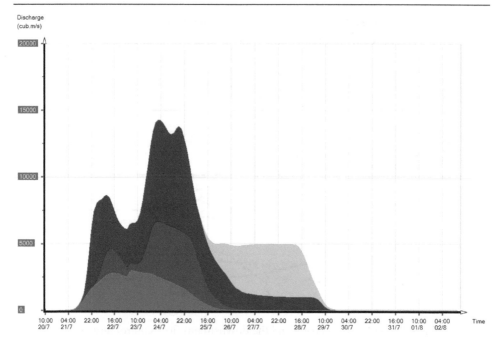

Fig. 3.3. Example hydrograph at Huayuankou station

Fig. 3.4 (A) and (B) show such maps. However, these maps are static and based on particular floods of some years ago and therefore may not be valid for present conditions. Due to frequent scouring and sedimentation process, the river bed and cross sections changes a lot.

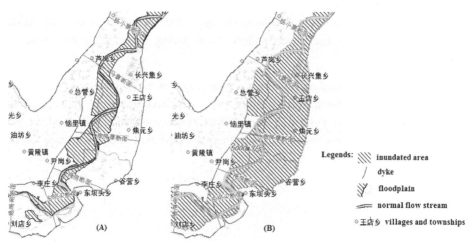

Fig. 3.4. Inundation maps made from historical flood events. Map (A) is based on flow 3000 m³/s, and (B), 6000 m³/s (valid before 2005) (source: YRCC)

Therefore, the possible flooding situation may be quite different from those historical event based maps. Moreover, the detailed information, for example, the distribution of inundation, is impossible with those maps. So, the conclusion is that 2D simulation model is needed to provide a means of flooding prediction tool for flood management and decision.

3.3 Review of multi-reservoir simulation tools

During the present research, some of the popular multi-reservoir tools were studied to determine the capabilities and application conditions of the existing software systems. Also the software system used in YRCC is examined to find out what should be improved for future successful systems.

In China, the ICT applications in the flood control and management field are popular now with a big variety and a fine division of different aspects. These applications include automatic gauging, data collecting and processing, historical and real-time hydrologic data retrieving, flood forecasting, flood control management and decision support.

The China's National Flood Control and Drought Relief Commanding System (NFCDRCS), which is organized by the National Flood Control and Drought Relief Headquarters and funded by the state, and is until now the biggest system of its kind and also a national wide program covering all the important gauging points and basins of the entire country. The NFCDRCS is aimed at using ICT to upgrade the flood gauging, data collecting and processing, flood forecasting, flood control and management hardware and software systems in China's main rivers and lakes systems, especially for the seven big basins, i.e. the Changjiang (Yangtse) River, the Yellow River, the Zhujiang (Perl) River, the Songliao River, the Haihe River, the Taihu Lake and the Huaihe River basins.

For flood forecasting, the so-called "China Flood Forecasting System" was developed and had been finished. The character of the system is that it carries out the forecasting of important rivers and region of the whole country with fully automatic daily or real-time routine, with each regional river authority having a subsystem linked to the national one. This means that the flood forecasting software, which is usually run to get forecast information when a user needs, is run automatically. Then, the user can access the forecasting results at any time, like retrieving any ordinary database. In case of a big flood, more specific forecasting will be made and issued.

In YRCC, the development of the flood control and management software system started in early 1980s. During the national "fifth five-year plan" of 1985-1990, the first software tool was developed to serve the joint operation of the three reservoirs in mid-lower Yellow River. This project was named the Yellow River Flood Control Decision Support

System. It was based on the DOS operating system and had some graphical interfaces. It was also a pioneer system of its kind in China. Therefore, a lot of basic research was done during the design phase of this project, including functional requirement analysis, decision support and system operational style. After the system was completed, it was put into real-time operation for several years, till late 1990s. Then, a Sino-Finland cooperation project was launched covering applications of flood forecasting, flood control, water quality simulation and water resources allocation management. Thus, a new version of the flood control system was developed on UNIX system with X-Window graphical interfaces. Not long after this system was finished, development of another PC-based system started. This was the flood forecasting and flood control coupling operation system, which served the real-time operation for several years, up till 2005 when the NFCDRCS project was launched.

The multi-reservoir flood control software system, under NFCDRCS, was designed in 2005 and then development was completed in 2010. This is the system being operated now. In general, this is a system developed based on the present systems and experiences. Therefore, progress can be seen if compared with the previous ones. However, there are still complaints about the computational kernel, user interfaces and decision support style. This is also a main driving force for this research.

Deltares is a Netherlands based research and development institute. Deltares river basin planning and management tool, RIBASIM (river basin simulation) is a generic model package for simulating the behaviour of river basins under various hydrological conditions. The model package is a comprehensive and flexible tool, which links the hydrological inputs at various locations with the specific water-users in the basin. RIBASIM enables the user to evaluate a variety of measures related to infrastructure, operational and demand management and to see the results in terms of water quantity, water quality and flow composition. RIBASIM can generate flow patterns which provide a basis for detailed water quality and sedimentation analyses in river reaches and reservoirs. RIBASIM provides the means to prepare water balances at a required level of detail, taking into account drainage from agriculture, discharges from industry and the downstream re-use of water. Typical applications include the evaluation of the options and potential for development of water resources in a river basin; the assessment of infrastructure, and operational and demand management measures (Deltares, 2012).

Van der Krogt (2012) mentions as applications of RIBASIM the projects in West Java (Indonesia), Eastern Nile (Egypt) and four basins in Morocco. These examples show the applications mainly for capacity building, spatial planning and policy development, water balance simulations, optimal integrated management and simulations for a number of identified scenarios, measures and strategies like water infrastructure development and climate change.

The HEC-ResSim software tool, developed by the Hydrologic Engineering Center of the US Army Corps of Engineers, is a modeling system that aid engineers and planners, performing water resources studies, in predicting the behavior of reservoirs and to help reservoir operators to plan releases in real-time, during day-to-day and emergency operations (USACE, 2011). ResSim is comprised of a graphical user interface (GUI), a computational program to simulate reservoir operation, data storage and management capabilities, and graphics and reporting facilities. ResSim offers three separate sets of functions called Modules that provide access to specific types of data within a watershed. These modules are watershed setup, reservoir network and simulation (USACE, 2007). Each module has a unique purpose and an associated set of functions accessible through menus, toolbars, and schematic elements. Just as its name implies, the ResSim is reservoir system simulation software tool. It is a powerful tool for river-reservoir system simulation and management.

MIKE BASIN, by DHI (Danish Hydraulic Institute) group, is a similar software tool that has reservoir modeling capabilities. It can accommodate multi-purpose and multiple reservoir systems. MIKE BASIN establishes dynamic water balances, including gains and losses, evaluates reservoir yields and investigates best operation policies, evaluates water banking opportunities through storage rights system. MikeBASIN represents complex operating policies by scripting (MikebyDHI, 2012), using Visual Basic macros to let the user define the operation policy. Reservoirs can be modeled with the methods of standard reservoir and allocation pool reservoir. For the reservoirs, the performance of specified operating policies using associated operating rule curves can be simulated. Rule curves define the desired storage volumes, water levels, and releases at any time as a function of existing water level, the time of the year, demand for water and possibly expected inflows. With standard reservoir method, all water users draw water from the same storage volume and operation rules regulate the water user's extraction from the storage pool. The allocation pool reservoir functionality is as follows: water is drawn from a storage volume and operation rules regulate the water user's extraction from the storage pool. However, the allocation pool method subdivides the storage by user's storage right (DHI, 2006). In addition to these two reservoir methods, there is the lake method which implies no operation rules, but a water level-dependent outflow.

Fig. 3.5 shows the general process related to the application of reservoir simulation software. The processing blocks above the dashed line are the tasks which are for the initial setting and configuration of the system. Those below the dashed line are the operations that users of a system needs in order to operate the system. From the user point of view, more effort should be put to the lower part. This part should be designed and implemented with good functions and user interfaces.

What has been shown in Fig. 3.5 is a process very much in the IPO (Input, Process and Output) operational or execution pattern. Namely, after the system is set up, every use of it starts with the input of data and setting of parameters, then runs the system, and then finally checks the output result and makes analysis. Most of the simulation system runs in this way. This is the case of HEC-ResSim, which gains its name from "Reservoir Simulation". So they are not typical management or DSS oriented tools.

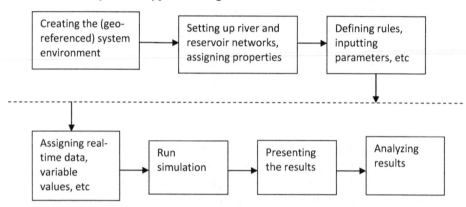

Fig. 3.5. General application procedure of multi-reservoir simulation system

3.4 Calculations required for YR reservoir operation

Adaptive management and decision support implies more requirements to the routing processes over the basic routing methods. Thus, more detailed calculation processes are needed to develop a real flexible and powerful tool. From the practice on Yellow River, it was found that the controlled release is a common regulation in analysis and alternative making in case of real-time flood control and management. The controlled release here means that there is gate open or closing operation during the routing process of a given time series of discharges. Chow (1964) stated that if the gates are operated during the routing period, (i.e. the controlled release operation), then the routing process becomes complex. However, the calculation methods or schemes can neither be seen much in literatures, nor in the technical references or manuals of popular software tools. The controlled release routing and regulation, which is required especially by adaptive management and decision making, is usually ignored and therefore absent for the simple reason that most of the reservoir management tools are simulation systems which are generally based on uncontrolled release operations. Problems related to the controlled release calculations have occurred many times in the Yellow River flood control software tools. It can be concluded that the controlled release has not drawn enough attention though it is a frequently requested regulation process.

In Yellow River flood control operation, the typical controlled release regulations may contain one or more of the followings: (i) release a planned hydrograph; (ii) keep a given storage level; (iii) different release policy for below, equal to or above a given level; (iv) constant release discharge for a given storage variation extent; (v) a simultaneous or an average release discharge for a time step is based on inflow or downstream conditions, such as values of inflow or downstream hydrograph at certain point. In additional to this, if different uncontrolled release scheme are used for a time series routing, such as different combination of opened sluices for various time periods, then the routing is also controlled release. All these controlled release regulation requires more complex computational scheme in comparison with the simple uncontrolled release. Traditional storage-indication method, the Plus method, is used for uncontrolled release routing and cannot be used for controlled release calculations. In some literatures, the Runge-Kutta method is suggested for storage routing, but it requires also the uncontrolled release operational condition.

It is also important to deal with the complex problem of transition from one uncontrolled or controlled release to another uncontrolled or controlled release, because the mass conservation has to be strictly kept and software tests in YRCC show that calculation for this transition needs to be improved.

As an example, in China, most of the reservoirs have their flood season normal level (FSNL). During the flood season, for flood control purpose, the storage level is not allowed to exceed the FSNL. On the other hand, the reservoir operation authorities will try to keep the actual operational level as high as possible because they always want to get the best benefit from power generation etc. Thus, the FSNL will be the right and unique level to be used as the general operational level during the flood season. In this case, the reservoir's release curve, the relation of storage level and release discharge will no longer be valid at this moment while the water quantity conservation should still be strictly kept. To meet this, the value of storage level has to be checked in every computational step (time interval). If the level is going to go below the FSNL, then controlled release calculation scheme is needed to make the storage level equal to FSNL while keeping the mass of in-pool water in strict conservation.

3.4.1 Reservoir routing conditions and routing methods

In general, the inflow into a reservoir is a series of discharges varying in time and can be in the form of a forecasted inflow, a designed flood inflow, etc. In case of real-time flood control and management, the input is usually the forecasted flood time series. The reservoir characteristics usually consist of discretized data of its physical features. The items that are mostly used for such descriptions are elevation (storage level) and the corresponding storage and release capacities of various outlet structures. Table 3.1 gives an example of such characteristics. Sometimes, the water surface area sizes are also given for different elevations. Release discharge and numbers of release facilities are given to each of the release structures. For the example, the bottom orifice release discharges and the number, 2, in Table 3.1.

Table 3.1 Example of reservoir storage and release characteristics

Outlet structures		Spillways	Middle orifices	Bottom orifices	Generator sluices
Total number		5	1	2	3
Storage level (m)	Storage (billion m³)	Release capacity (m³/s)			
478	0.015	0	0	80	0
480	0.048	0	0	106	0
490	0.114	0	0	200.5	0
500	0.196	0	0	276	36
510	0.295	0	0	329.5	36
520	0.412	0	0	375.5	36
530	0.561	0	1122	416.5	36
535	0.658	132	1266	435.3	36
540	0.773	558	1303	453.5	36
550	1.08	2029	1461	488	36
551	1.167	2287	1476	491.5	36

Reservoir characteristics data is initially made through field survey followed by calculations during the design phase of a dam. If the reservoir bed is static or changes very little, then the design characteristics data can be used for long time with no need of updates. However, the data has to be updated periodically with reservoirs which have substantial bed changes due to sedimentation and scouring. The Xiaolangdi and Sanmenxia reservoirs, which are used in this research, belong to the second type of reservoirs. Thus, their characteristics data is updated two times a year, one before and another one after the flood season. In the Yellow River the flood season is usually within the time period from 1 July to 31 October.

Modern reservoir-based flow management becomes more and more complex in terms of operational schemes and objectives. An example on the Yellow River is the so-called multi-reservoir-based flow and sediment transport coupling regulation which aims at regulating the flow and sediment load to a favorite ratio so as to transport as much as possible sediment to the sea (Li, 2009). In the operational scenario, up to 20 days a release plan is made for all reservoirs taking part in the regulation operation. Each reservoir has a detailed release scenario with given release discharges or opening modes for every time step of the operation. The release discharges can be either instantaneous values or time interval averaged values. The opening modes can be changed many times during the entire operational time span. Thus, this coupling regulation between flow and sediment load demands very complex reservoir routing processing.

It is necessary to have a clear classification of different types of routing conditions so as to have a good road map for solving the problem. As shown in Fig. 3.6, the reservoir routing conditions can be divided into two types: uncontrolled and controlled release operations. These two types can be further divided into sub-types, as follows:

(1) Uncontrolled release

Uncontrolled release means the release structures keep their operational modes unchanged during the entire time series routing process. The uncontrolled release conditions includes: type (i) **release with smooth storage level (H) - release (q) relation**, i.e. a continuous H-q curve, or a functional relation; type (ii) **release with non-functional H-q relation**, i.e. with a discontinuous or multi-valued H-q curve.

Type (i) has a good storage level-release (H-q) relationship or smooth functional relationship. Every value of q has only one corresponding storage level or storage. Type (ii) is uncontrolled release operation but with bad H-q relationship, for there is sudden sharp increase of q at a value of storage or storage level. This usually happens with the joining of a release structure at certain elevation, which has much bigger release capacity, like spillways for extreme floods. The sharp increase may be the bottom of the spillways.

(1) Controlled release

Controlled release is more complex operation. It can be further divided into four sub types: type (iii) **release with given H-q relation**. This H-q relation may consist of both uncontrolled release graphical sections and controlled release graphical sections, for example, the vertical straight line section where the release discharges remain constant and do not change. So, it is a 'controlled' release; type (iv) **release with given release time series**. In this case, the release is given. So release discharge will not be decided by storage or storage level. But a given discharge number at certain moment has to be smaller than the actual maximum release at that moment.

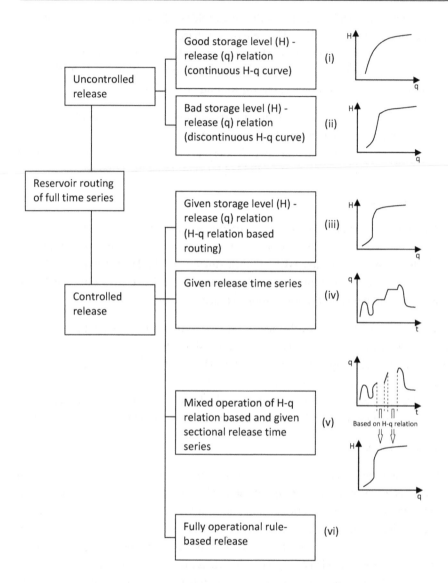

Fig. 3.6. Different conditions for reservoir routing

Type (v) **release with mixed operation given sectional release time series and H-q relation based operation**. This is the most complex process. For the entire time series of release, some time periods of release discharges are given while other sections are based on one or more given H-q relations. These relations may be uncontrolled release or controlled release; (vi) **fully operational rule-based release.** This type refers to the release operation that the actual release discharge is controlled by flood control and management rules. For example, the release discharge is decided by the upstream inflow discharge or the downstream discharge at certain gauging points. Here the word 'fully' is used because the other types may be partially controlled by some rules. An example is type (iii). The given constant release section on the graph is the operation based on rule(s). Fig. 3.7, Fig. 3.8 and Fig. 3.9 are three examples with different release types.

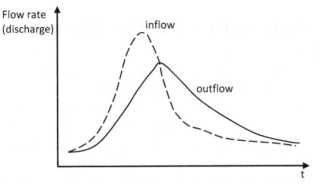

Fig. 3.7. Sample hydrographs for uncontrolled release

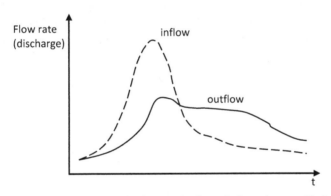

Fig. 3.8. Sample hydrographs for rule-based controlled release

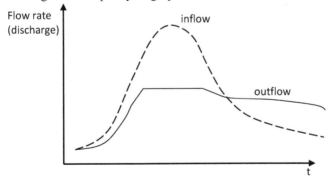

Fig. 3.9. Sample hydrographs for controlled release with fixed maximum discharge

Adaptive management requires knowledge on the inflow and outflow conditions. Since traditional routing methods generally assume that the release is a single-valued function of storage or elevation, then they are not able to deal with all condition types listed above. So, a new method or algorithm needs to be developed.

3.4.2 Routing method requirements for fixed H-q relations

Uncontrolled release typically has two types (see (i)and (ii) in Fig. 3.6). Hence, the new method should also work for these types, to extend the basic routing method to a wider range of application conditions. In Fig. 3.10 the curve q_1 is a regular one with a quite smooth behaviour; q_2 has a vertical line in its middle section, which corresponds to a H-q relationship based on operational rules, rather than on uncontrolled release. So it is a given fixed H-q relation with controlled operation of release facilities. Curve q3 starts from small storage level with bigger release value and goes towards smaller values in the low level section. Although q_3 looks different from what would be expected it is a reality for some reservoirs. For example, the Sanmenxia reservoir generators' release curve will become smaller as storage level goes higher. If the release is only based on generators' outlet (one or more generators), the storage level and release relation will be in the same pattern of the lower section of curve q_3. All three have a sharp increase sections at high storage levels. This is normal because high storage levels, above a certain threshold, require big releases, for dam safety.

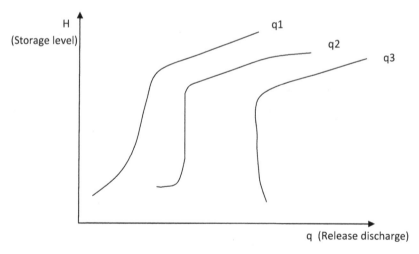

Fig. 3.10. Example of different types of storage levels and corresponding releases

As stated above, the general reservoir routing methods assume the release q is a function of storage level or storage volume. This condition is right with q_1 curve, but not with curve q_2 and q_3. Mathematically, curve q_2 shows a relation of storage level and release, not a mono-valued relation. And so is the q_3 curve. In case a functional relationship is used for routing purposes, calculation errors and problems can occur. Therefore better routing approaches and/or methods should be explored. These will be presented and discussed in Chapter 5.

3.4.3 Requirements for controlled release routing

Complex management rules or scenarios lead to complex routing calculations. The typical example is maintaining a given target level during operation. In this section the requirements are explored and explained.

Regulation for keeping a given target level is one important regulation process which is usually needed with the flood control and management of reservoirs when the "flood season normal level" needs to be maintained. During flood season, as long as there is no flood happening, a reservoir is required to be operated at a level equal to, or lower than the flood season normal level.

At the same time, during flood time, the water level is required to be maintained at this level, so that it can meet the requirements of later floods. In most cases, the reservoir operator may also want to keep the operational level as high as possible for power generation. Thus, the actual level is usually kept at the flood season normal level.

The regulation for keeping a given target level is a complex rule-based controlled release routing process. When the level is below the given target level, it is generally required to store as much water as possible, so that the target level is reached as soon as possible, for power generation purposes.

When the level is above the given level, releases are made, in order to return to the flood regulation function. So, if the level is below given level, the release is set to be either zero or a minimum value. If the level is higher than the target level, the reservoir could be in operation of a flood regulation process. The release options include all-open release, partly-open release and given release hydrograph.

Strictly speaking, there are up to seven different patterns of level changes around the given target level. They are shown below in Fig. 3.11.

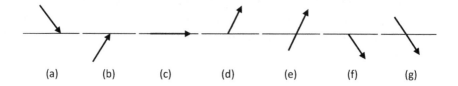

Fig. 3.11. Different patterns of level changes, with respect to a given target level
(the horizontal line)

The different symbols for level changes represented in Fig.3.11 are:

(a) Reach target level by releasing water (actual water level is above target level);

(b) Reach target level by storing water (actual water level is below target level);

(c) Keep the given target level;

(d) Increase from the given target level to a higher water level;

(e) Increase from water level below the given target level to a water level above;

(f) Decrease from the given target level to a lower water level;

(g) Decrease from water level above the given target level to a water level below;

The last two cases are not usually occurring during normal reservoir operation, they are special cases that do not happen so often. Two situations could be:

(i)   when flushing the sediment from the reservoir bed, and

(ii)   when extending the flood regulation zone, by adding part of the storage below the flood season normal level to the standard flood regulation zone.

If the level is to go below the flood season normal level, other computational components will be used, including all-open, partly-open or given-release time series components based on actual operational conditions.

Keeping a target level involves controlled releases. If the process is looked into in a smaller scale, it will be found that the given target level is most probably reached at certain moment within a time step interval. The instantaneous values at both ends of the time interval are correct and will be a result of controlled release. However, the change of release from the beginning of the time step to the end of it will most unlikely to be in linear change. Fig. 3.12 gives an example.

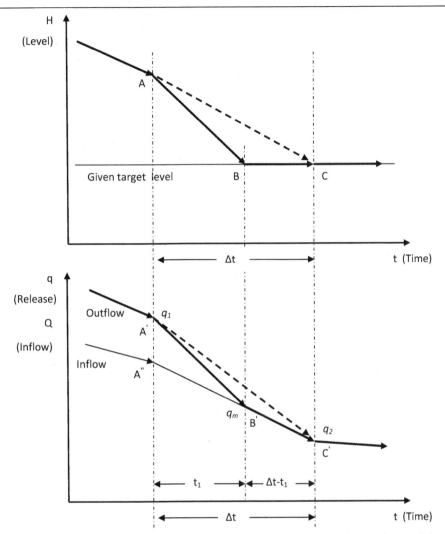

Fig. 3.12. Example: the given target level is reached within the time interval Δt and then kept at that level.

In the example given in Fig. 3.12, the level is higher than the given target level, at the beginning of the calculation time step, therefore the outflow is bigger than the inflow, in order to lower the level towards the target level. Once the target level is reached (point B and B'), the outflow will be controlled so that it is equal to the inflow (from point B and B'). Therefore, the given target level is kept. What makes it different for this time interval, from most of others, is that the instantaneous value of release at both ends is the change from point A' to B' and then from B' to C' rather than directly changing from point A' to C'. Within the time interval, the regulation calculation has changed from previous mode, uncontrolled or controlled release mode, to a new controlled release mode, the mode of

57

releasing given time series (the inflow). This needs special processing for the reason that, in general, the two instantaneous values of inflow at both end of the time step are used to present the release, but the two values cannot precisely give the real release. The change between the two ends of the time step cannot be reflected by the two instantaneous values.

The main steps of computing the releases are exemplified bellow, as follows: let q_t, q_{t+1}, Q_t, Q_{t+1} be equal to the instantaneous values of release and inflow discharges for time step t at the starting and ending point of the computation (marked with t and t+1 respectively). The term $\frac{q_1+q_2}{2}$ could be quite different from $\bar{q}$, which is usually used in routing calculation. That is:

$$\bar{q} \neq \frac{q_1 + q_2}{2}$$

In the example shown in Fig. 3.12, it can be seen that the average release over the time interval Δt, should be in the relation:

$$\bar{q} < \frac{q_1 + q_2}{2}$$

However, the values at both ends of the time step are real and correct instantaneous values for those two time points.

Let the time from point A' to B' equal to t_1, then the time from point B' to C' is Δt-t_1. At same time, let q_m equal to the release discharge at point B'.

The average release over the time interval Δt can be calculated with:

$$\bar{q} = \frac{\frac{q_1 + q_m}{2}t_1 + \frac{q_m + q_2}{2}(\Delta t - t_1)}{\Delta t}$$

From Fig. 3.12 we know q_m is the instantaneous release discharge at the moment when the storage level reaches the given target level. It is also the instantaneous inflow of that moment. This instantaneous inflow is not given, because it is in the range between two given inflow numbers, Q_1 and Q_2 at both ends of the time step. Thus, it can be calculated through linear interpolation with Q_1 or Q_2.

Since both q_m and t_1 are unknown, $\bar{q}$ cannot be calculated directly. If $\bar{q}$ or its location, t_1 have to be obtained, one simple way can be an option: use a smaller time step (a smeller Δt) to make more delicate routing process and, at the same time, mark the moment that the target level is reached. Then, $\bar{q}$ of the original time step can be calculated with the average of all the smaller time step release discharges.

One typical unskillful calculation about keeping a target level is the level oscillation around the given target level. In this case, the calculation is not made perfectly. Much of the effort here is to obtain a smooth routing and resultant level variation process. Based on the Yellow River flood control practice, the operational modes above target level have three options: all-open release (release in capacity), partly open (with some release structures opened) release and given release limitation process. Below the target level, the underlying principle is to get back to the target level as soon as possible, so the operation is either the uncontrolled release of the given minimum release or other given release.

3.5 Reservoir operation and downstream safety

In the early stage of the present research, the Yiluo branch (Fig. 3.13) has been selected to build a SOBEK 1D2D simulation model to study the effect of flood control by reservoirs (Tang et al, 2009). The Yiluo branch of the Yellow River has the two reservoirs, Guxian and Luhun, on the sub-branches, Yihe and Luohe, respectively. In the model, three aspects of SOBEK were implemented, due to geography and the application of different module requirements: for the upstream of the two reservoirs on Luohe river and Yihe river, only rainfall runoff (RR) module was applied since both of the upstream sections of these two rivers located in mountain area, the majority of flows were generated by rainfall. For the middle and lower reach of rivers, 1D channel connected with Sacramento model was applied in order to obtain rainfall runoff lateral flows. And considering the observation of flood maps under some storm events, 2D module was applied in a downstream area where there has been dyke breach and overland flow in the flood of August 1982.

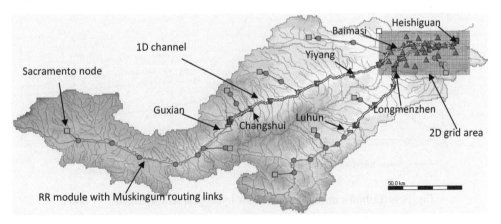

Fig. 3. 13 Application of SOBEK RR/1D2D model with the Yiluo branch

In order to see the flood regulation effect clearly, a historical rainfall event from August 1982 was used to test the function of multi-reservoir-based operation in the model area. This event is approximately of a 0.1% frequency flood magnitude for Guxian reservoir, in terms of inflow peak and volume.

During the study, four scenarios were calculated. The main differences between scenarios were the openings of the sluices and the initial storage levels of the reservoirs. Namely, the release facilities are either all opened to have maximum release or partly opened to have a controlled release. The initial storage levels can be the flood season normal level (FSNL) or a lowered level which means pre-release operation is made for regulating extreme floods. The four scenarios are: (i) maximum release with initial storage level FSNLs; (ii) optimized release with initial storage level FSNLs; (iii) maximum release with lowered initial storage levels; and (iv) optimized release with lowered initial storage levels. Fig. 3.14 shows the Luhun reservoir's hydrographs for inflow and two outflows of Scenario 1 and 2. It can be clearly seen that Scenario 2 regulates the flow in a much bigger extent. The outflow discharges become much smaller. In this case, the discharging of flood water storage during the regulation process needs a longer time. But this scenario can keep the downstream area safe. The basic condition for this operation is the safety of the dam.

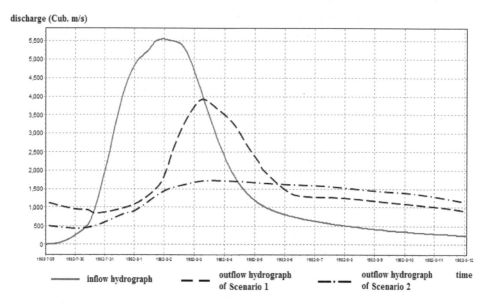

Fig. 3.14 Luhun's inflow and outflow hydrographs for Scenario 1 and 2

With the four scenarios, the initial flooding time in the 2D low land is obtained and then comparisons are made: from scenario 1 to scenario 4. The times for comparisons are: (1) 12:00 1/8/1982; (2) 16:00 1/8/1982; (3) 16:00 1/8/1982 and (4) 02:00 2/8/1982, at the moment of the maximum flooding extents for each of the cases. It can be seen that the flood firstly occurs along Luohe river for all four scenarios, but the starting time for flooding is different. Regulated flow helps in delaying the starting time of the flood. The worst scenario (scenario 1) is the first to have flooding which appears half day quicker than that in the best scenario (scenario 4). Fig. 3.15 through Fig. 3.18 are given for the comparison of maximum flooding depths in the 2D low land in the four scenarios. The figures were captured at the time when flooding depths are seen to be the biggest.

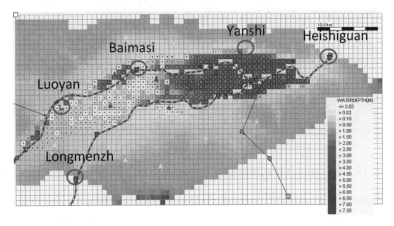

Fig. 3.15 Scenario 1: time of largest flood depth occurrence, at 10:00 4/8/1982

Under scenario 2, most parts of Luoyang City (see Fig. 3. 19) sinks in water with a maximum flood depth around 2 meters, while part of Yanshi City (see Fig. 3. 19) was also affected by the flood.

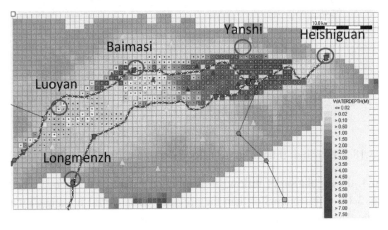

Fig. 3.16 Scenario 2: time of largest flood depth occurrence, at 20:00 4/8/1982

Due to the operation of the Guxian reservoir, the flooded area represented in Fig. 3.16 was smaller (about one half) than in scenario 1. Luoyang City was still flooded, but the maximum water depth was less than 0.5 meters. Yanshi City was threatened.

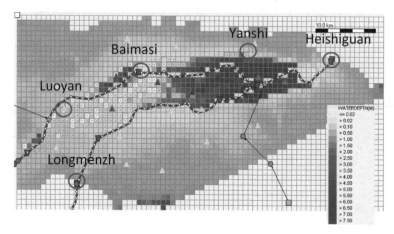

Fig. 3.17 Scenario 3: time of largest flood depth occurrence, at 18:00 4/8/1982

Under scenario 3, Luoyang City got a little flooding but not very serious. The southwest corner of Yanshi City got flooded with 1 meter water depth in maximum.

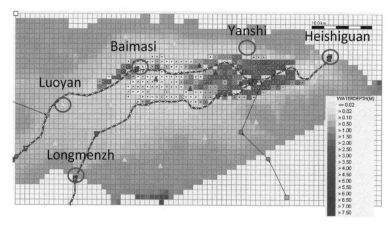

Fig. 3.18 Scenario 4: time of largest flood depth occurrence, at 00:00 5/8/1982

In scenario 4, the regulation of inflow gives a bigger extent of the flood. The total flood area, in Fig. 3.18, is much smaller than the previous ones. Both Luoyang City and Yanshi City are safe. The inundation mostly happened in rural area of small towns with lower flood depth and velocities than previous scenarios so that less people's lives were threatened by this storm event, the economic loss could also be reduced.

Fig. 3.19 Location of Luoyang City and Yanshi City (map source: Google Earth)

Scenario's 2, 3 and 4 were based on and evolved from scenario 1 by changing the operation modes. It can be concluded that the reservoir regulation of floods plays a very important role in reducing flooding threat downstream, if the inflows into the reservoirs take into account the entire flood extent of the area. Different scenarios should be developed with various considerations, guided by general rules to obtain the information that can provide a comparison of different operational measures to make a good decision.

3.6 Summary

From the practice on the Yellow River, it was found that the so-called 'controlled release' strategy is a very common regulation principle for real-time flood control and management. Controlled release has not drawn enough attention though it is a frequently requested regulation process. The multi-reservoir regulation on the Yellow River demands not only uncontrolled but also controlled release schemes, including constant release for water level variation zones.

Moreover, different zones may have quite different storage-release relationships which require special attention and routing techniques so as to have a correct and robust routing process. The controlled release regulation requires a more complex computational scheme in comparison with the simple uncontrolled release.

It is also important to deal delicately with the transition from one uncontrolled or controlled release to another uncontrolled or controlled release, because mass conservation has to be strictly kept and software tests in YRCC show that calculations for this transition often exhibit oscillating behaviour which needs to be improved.

Complex management rules or scenarios lead to complex routing calculations. The typical example is the maintaining a given target level regulation operation. This regulation calculation is a complex rule-based controlled-release routing process. The traditional methods do not fit this calculation. Oscillations around the target level have repeatedly been observed in some software tools. Therefore, a robust algorithm and program implementation is needed.

4 Numerical flood simulation on lower Yellow River

Chapter 4 presents the studies regarding the propagation of floods on the Lower Yellow River reach using a 1D2D approach. Simulation models are basic tools to evaluate the impact of different reservoir operation alternatives. Different ways to operate the reservoir system of the Yellow River may have different inundation effects on the lower Yellow River floodplains. The prediction of the inundation pattern on the Yellow River was done using the state of the art simulation tools produced by Deltares; SOBEK Rural and the new D-Flow FM *beta* in their 1D2D set-up. The two tools use different solution approaches, a uniform grid and a flexible grid, respectively, in order to build the simulation models, which can be used for flood management and decision support. Discussion on which of the two approaches is better suited for a decision-support is discussed and evaluated.

4.1 Introduction

The lower Yellow River is defined based on the presence of formal dykes on both sides of the river. The starting point of dykes is at the upstream boundary of the lower Yellow River reach. The total length of the lower Yellow River is 786 km. This part of the river is also known as a "hanging river", because its reach is a suspended channel, not a part of a normal leaf-shaped catchment. Moreover, more recently, a special feature is that within the channel, the low flow route, where normal flow (characterized as non-flood time flow or dominate flow for most time of year in non-flood season) goes hanging as well.

The floodplains of the lower Yellow River cover two provinces, Henan and Shandong. The total area of the floodplains is about 4000 km^2 with a number of 1.9 million people living and farming in this area (Yan, 2004; YRCC, 2009). The Yellow River has many special features caused by its heavy sediment transport. Sedimentation takes place since a very long time and as a consequence nowadays the channel is higher than the floodplains, therefore a smaller hanging river develops in the hanging Yellow River channel. Fig. 4.1 depicts a cross-section of the Yellow River that reflects the hanging river pattern of it.

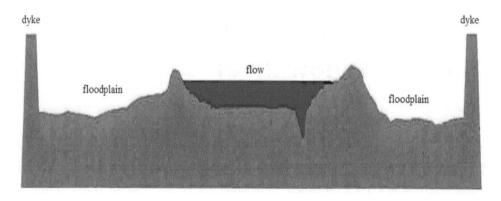

Fig. 4.1. Typical cross-section of the of lower Yellow River channel (source: YRCC)

Once the flow goes out of the main channel, it will go to the lower areas of the nearby floodplains.

Because of fast socio-economic development in the last decades, the infrastructures have been greatly improved to a higher level in the floodplains, with a beneficial impact on people's life. In the same time, higher expectations and standards were imposed from safety point of view, as well as from sustainable development of these areas. Decades ago when these floodplains were frequently hit by floods, there was little that the river authority could do, simply because the controlling engineering system was not enough. Nowadays, conditions have changed a lot. More reservoirs have been built in the upper part of the Yellow River reach with the aim to control flood to the downstream.. Flooding caused by natural flow happened more frequently in the past, but now, flooding, if any, would most probably be a planned action and uses the floodplains to detain the inundation.. This change makes the flood control operation essential and critical. Operational schemes have to be made more sophisticated and accurate. This makes the flood control management a great challenge.

Throughout time many efforts have been done, by the Yellow River Conservancy Commission, the river's management authority, in developping simulation tools, which will help achieving the goal of having a good management of the system and to control the flooding in the area. The so-called "YRCC Model", which is a coupled 2D flow and sediment transport model for the lower Yellow River channel (YRCC, 2006b), has been now used for several years and has been tested under real operational conditions. The validation of the simulations of the YRCC Model was always a problem because of the lack of fairly accurate floodplain inundation maps. However, as a substitution, flood maps based on actual flood evens, for which the return period of the flood has been statistically determined, might be used to validate a model. These maps were made for different historical flood events, for example 3000 m^3/s, 4000 m^3/s, etc. Fig. 4.2 shows an example of the Yellow River flooding for a 3000 m^3/s discharge, which is a map based on historical events before 2005. However, these maps are valid only for a limited time, because the lower Yellow River bed and terrain changes frequently.

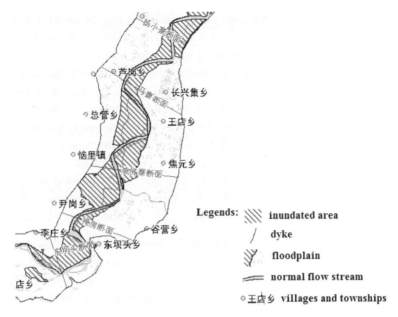

Fig. 4.2. Example map of inundation in the Lower Yellow River near Jiahetan hydrologic station (source: YRCC)

In this PhD research two types of models were used in order to model the flood extent and to see the effect of the reservoir operation on the flood. The two models are based on the software that was developed at Deltares, in The Netherland, and which has been tested in many flooding situations throughout the world. The two model systems are using structured and non-structured grids, respectively. The non-structured grid approach is a new and innovative way of approximating floods, and has been tested for accuracy, speed and easiness of being used for complex cases. The model of the lower Yellow River reach, based on a structured grid, uses the SOBEK modeling suite and the model based on the unstructured grid uses the newly released DFlow beta. The two models are built with the aim to predict the possible inundation area of operational alternatives in an effort to link the multi-reservoir operation to their possible influences on lower Yellow River floodplains. The models use hydrograph of the Huayuankou hydrologic station (point A on Fig. 2.3) as upstream boundary condition. In real-time operation, the hydrograph is given by forecast with certain lead time.

4.2 Numerical modeling of flood events

The principles of simulating flood in rivers are the conservation of volume, momentum and energy. These equations were developed centuries ago by Newton (1687), who introduced them in a clear mathematical formulation, based on physical conservation principles. Using these conservation principles, later on, De Saint Venant (1871) formulated the mathematical equations for modern day river flow simulation. However, at the moment of their formulation no solution was available. It required the development of nowadays powerful computer in order to obtain suitable solutions of these equations, by using numerical techniques to solve them for practical applications (Stelling and Verwey, 2005). The numerical flow simulations become practical nowadays due to two main developments in the past decades: one is the development of practical numerical schemes for 1D, 2D or 3D computations; and the other one is the development of the fast computers (Balica et al, 2013; Van et al, 2012). These two aspects stimulated and helped each other and led to the development of widely applicable and practical tools.

The four software giants in the field, Deltares, The Netherlands; DHI (Danish Hydraulics Institute); US Army Corps of Engineers and Innovyze (former Wallingford Software), all have developed their own flow simulation tools. Of all these tools, the Deltares suite, such as SOBEK 1D2D and DFlow beta, which integrates the simulation model for the 1D river course and the 2D overland flow (Moya Quiroga et al, 2013), is a good choice for the lower Yellow River flow and evaluation of the flooding in the area.

This section of the thesis makes a brief summary of the theoretical background of the 1D2D modeling approach.

Based on Fig. 4.3, the 1D Saint Venant continuity equation can be expressed as:

$$\frac{\partial A_f}{\partial t} + \frac{\partial Q}{\partial x} = q \tag{4.1}$$

Where:

A_f, wetted area (m^2);

Q, discharge (m^3/s);

q, lateral discharge to the channel per unit length and time (m^2/s);

t, time(s);

x, distance in the one dimensional x direction (m).

Physically, the first term in equation (4.1) represents the rate of change in water volume stored in a unit length of channel; the second term is the rate at which the discharge changes along the channel per unit time. The two changes together hold the conservation of mass. See Fig. 4.3 for definition sketch.

The 1D momentum equation reads:

$$\frac{\partial Q}{\partial t} + \frac{\partial}{\partial x}\left(\frac{Q^2}{A_f}\right) + gA_f\frac{\partial \zeta}{\partial x} + \frac{gQ|Q|}{C^2RA_f} = 0 \tag{4.2}$$

in which:

g, gravitational acceleration (m/s^2, around 9.81);

Q, discharge (m^3/s);

t, time(s);

x, distance in one dimensional x direction (m);

A_f, wetted area (m^2);

ζ, water level (m, with respect to the reference level);

C, Chézy coefficient (m$^{1/2}$/s);

R, hydraulic radius (m);

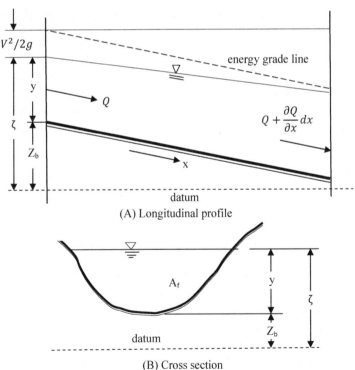

(A) Longitudinal profile

(B) Cross section

Fig. 4.3. Longitudinal profile and cross section of a 1D control volume of a river

At close examination of the terms of equation 4.2, from physical meaning point of view, we can conclude that the first term describes the inertia; the second one convection; the third one the pressure gradient; and the fourth one is the bed friction.

Numerical schemes and approximate solutions play essential roles in putting forward the use of the partial differential equations. It was not until Stoker (1957) that approximate numerical solutions were obtained, using an explicit finite difference method. Numerical solutions for differential equations have drawn the attentions of many researchers whose efforts were to solve them and put them into practical uses. A great number of different numerical schemes have been proposed.

In solving the problem, both uniform grids and flexible grids are used to discretize the simulation domain. With uniform grids, the domain is made with cells having exactly the same size of square or rectangle cell. The flexible grids allow various sizes of cells in the domain. This helps focus more on the representation of the flow fields based on different flow conditions and also allows the reduction of total number of cells (Dinh et al, 2012; Gichamo et al, 2012; Popescu et al, 2010).

4.3 The Lower Yellow River flood computational domain

The Shuttle Radar Topography Mission (SRTM) provides elevation data on a near-global scale to generate the most complete high-resolution digital topographic database of Earth. SRTM consisted of a specially modified radar system that flew onboard the Space Shuttle Endeavour during an 11-day mission in February of 2000. There are three resolution outputs maps available from the SRTM mission, the 1 kilometer and 90 meter resolutions for the world, and a 30 meter resolution for US.

For the research area of the mid-lower Yellow River, the 90 x 90 meter resolution data is the available one in the SRTM. This area covers a range approximately from 1100E to 1180E and $33^{0}30'N$ to $38^{0}30'N$. The corresponding SRTM DEM data sets are 58-5, 58-6, 59-5, 59-6, 60-5 and 60-6 (see Fig. 4.4)., The six image data sets that cover the specified coordinates were downloaded from www.ambiotek.com/srtm website.

Several steps of GIS processing with ArcGIS 9.3 were made to get the basic maps needed for the 2D representation of the floodplains. After processing the downloaded SRTM data the following elements could be extracted from it: (i) the river channel networks for building the 1D and 1D2D models; (ii) the catchment and sub-catchment extent; (iii) the DEM data for producing 2D grid maps; (iv) pour points, outlets of the watersheds, for joint points of 1D simulation model; (v) boundary information (ArcGIS shape files) for determining the limits of the 2D grids.

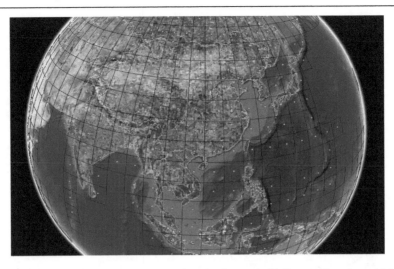

Fig. 4.4. Location of the SRTM DEM data (red trapezoidal) covering the mid-lower
Yellow River (source: Google Earth)

The map of the study area, as obtained at the end of processing the SRTM data, is shown
in Fig. 4.6. The upper part of the catchment, enclosed in a dark blue line, is a sub-
catchment in the middle reach of the river. This area is an important rainfall-runoff area,
contributing to the mid-lower floods, which are in turn threats to safety in the downstream
of the Yellow River. The four reservoirs under study are located in this upper part of the
catchment. In the lower part, the area enclosed by green line, the lower Yellow River
channel (area between the lower Yellow River dykes) is the hanging part of the river. In
Fig. 4.6, the dyke lines were drawn from other maps, available from YRCC, they are not
processed from the SRTM data. For most of the land outside the lower Yellow River
dykes, flows will not be able to go back and/or into the river, therefore the areas just
outside of the river's dykes are no longer contributing as rainfall-runoff catchment to the
Yellow River. This particularity makes the lower Yellow River roughly a tunnel which
simply conveys flow from its upper stream to its downstream and eventually to the sea.

In Fig. 4.6, mid-lower Yellow River was shown with the flood control and management
infrastructures. Due to China's high population density, the number of people living in the
floodplains of the channel is almost 1.9 million, while the number of people living in the
flood detention areas is 1.69 million. The prediction and evaluation of possible inundation
patterns of these areas are important to the flood control and management operations.

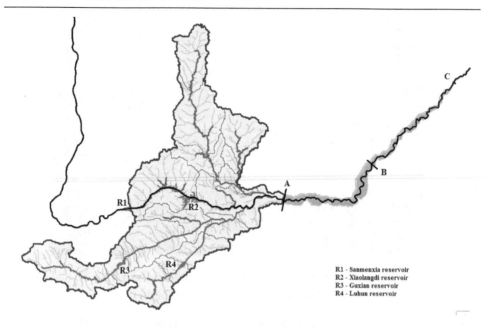

R1 - Sanmenxia reservoir
R2 - Xiaolangdi reservoir
R3 - Guxian reservoir
R4 - Luhun reservoir

Fig. 4.6. The position of the reservoirs in the mid-lower Yellow River

In Fig. 4.6, the start of the hanging river course is point A, where there is one hydrologic station, the Huayuankou station. This station is the key reference station for the reservoir operation, for the reason that it locates at a river section where the river changes from below ground to a "hanging river". In another word, from here the river becomes hanging. Point B in Fig. 2.6 is in the upstream of the intake gates of the Beijindi flood detention area. The hydrograph at point B (or Gaocun hydrologic station) is used for determining, in the case of extreme flood conditions, whether the Beijingdi detention should be used or not. Similarly, point C in Fig. 2.6, is just upstream of the Dongpinghu flood detention area, which is supposed to be used with the occurrence of big to extreme flood events. In some cases, the use of both areas is decided simultaneously based on situation of the entire mid-lower Yellow River. As previously mentioned, the use of the two flood detention areas is not dealt with in this research.

The operation of reservoirs is largely based on the hydrograph of Huayuankou station while the most concerned downstream impact will be the flooding of floodplains. If even after the regulation of flow by reservoir operation, an extreme flood event bigger than the standard dyke design occurs, then the flood detention areas are used. This shows that in most of the cases, the possible flooding and its extent are the main concern for flood control operation. Fig. 4.7 shows the major "wide channel" section of the lower Yellow River where most of the floodplains are and where there are 2 million people living. Furthermore, situation in the upper part of the catchment (upstream of point B) is crucial for flood control decisions. Due to the complexity of the problem and the way the two flood detention areas are operated the simulation of the downstream part of the YR channel conducted in the present study will not include the two detention areas The selected simulation domain for floods is shown in Fig. 4.8,

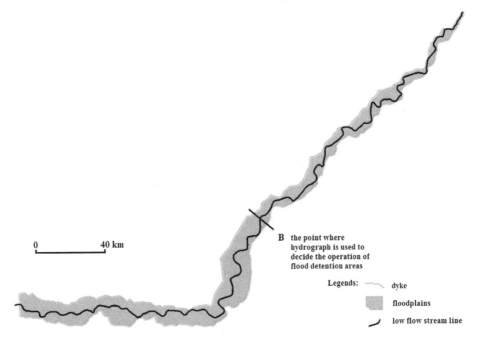

Fig. 4.7. The enclosure dykes of the lower Yellow River channel

The 1D river representation in the models is made based on the (low) flow route, which can be distinguished from the DEM map, as shown in Fig. 4.7. This makes the 1D river line sections more coherent to the natural states.

From the legend in Fig. 4.8, it can be seen that the selected area for the uniform grid is a square cell of 300 x 300 m, which is a big discretization cell. The DEM data from SRTM source, covering the Yellow River basin, is 90 m X 90 m. The width between the left and right dykes ranges from 5 km to 15 km. If 10 km is taken as the average and 150 km as the length of the river section, then the approximate number of computational cells covering the domain is:

$$\frac{10 \times 1000}{90} \times \frac{150 \times 1000}{90} \approx 185185 \text{ (cells)}$$

(Length unit: meter)

This is a long wide river and big floodplains in its downstream, therefore a big number of computational cells will require a very long simulation time, which is not desirable. Moreover, in case of 1D2D simulations the coupling requires the 1D river width to be smaller than the cell size. The nature of lower Yellow River does not allow for a small computational cell size. Thus, for all these reasons, the cell size of 300 m is used. Then the actual number of computational cells is approximately:

$$\frac{10\times1000}{300} \times \frac{150\times1000}{300} \approx 16667 \text{ (cells)}$$

(length unit: meter)

The value is eleventh times smaller than the previous one. In this case the time for one simulation proved to be acceptable , i.e. the simulation time is around 5-10 minutes on a PC with Intel i5 processer and 4 GB of memory. The big increase of cell size may lead to accuracy problems. However, for real-time operation purposes, if the computation time is too long (e.g. longer than the management meeting), then the simulation is useless, no matter it is more accurate or not. At the same time, even 90 m X 90 m DEM cannot be accurate enough for presenting the lower Yellow River terrain. For example the river training works, located between the normal flow course and the floodplains, are generally having a width of 10-20 meters, which is much less than a 90 m cell size. The purpose of the modeling in the present study was to show the principle of using such models, while for practical situations a more accurate model can be build, based on a more accurate DEM.

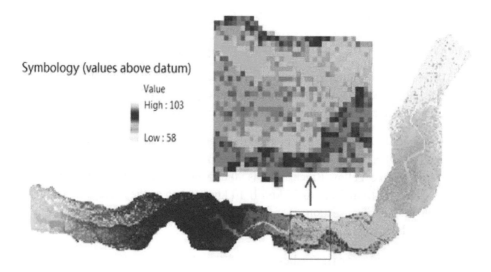

Fig. 4.8. DEM of the mid-lower Yellow River based on 300 m x 300 m cell discretisation.

4.4 Numerical modeling of YR using a uniform grid (SOBEK)

In this research, a model based on a uniform grid (square cell discretization) was built, using the SOBEK 1D2D. The uniform grids discretization model is shown in Fig. 4.9. The computational domain is spilt into a 1D network and a 2D system with rectangular computational cells for the floodplains. The 1D network and 2D system are forming a system of equations that is coupled and solved simultaneously with an implicit method (Dhondia and Stelling, 2002). In SOBEK 1D2D the water movement in the stream channel is solved by a finite difference approximation based on a staggered grid approach (Frank et al. 2001, Dhondia and Stelling, 2002). The 1D and the 2D schematizations are combined into a shared continuity equation at the grid points where water levels are defined.

The SOBEK 1D model set-up uses trapezoidal cross section type, with the upper width of the cross section of 290 m (<300 m); friction, Chézy 50 $m^{1/2}$/s; difference between surface level and the bottom level: 7-15 meters.

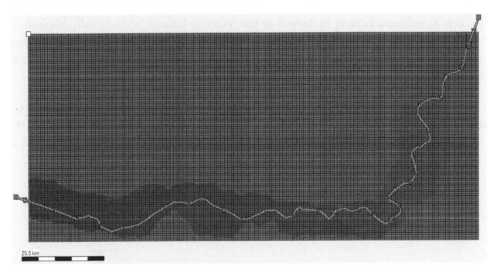

Fig. 4.9 The uniform grid of the SOBEK 1D2D model

The lower Yellow River floodplains have been very vulnerable to big floods and frequently hit by heavy flooding in history, but that didn't bring us much reliable flood risk maps for the newest lower Yellow River. One of main reasons for the lack of flooding maps is that the Yellow River course changes greatly and frequently in time. Thus, the post-flood mapping of flooded areas of the historical floods would not help much for prediction of today's new conditions. Due to the lack of detailed inundation maps of different floods for the latest river bed conditions, it is impossible to accurately verify the 2D simulation based on precise inundation maps. A simplified and yet practical way is to use the bankfull discharges (some critical values for flooding of floodplains) to calibrate the model based on the basic information of floodplain flooding. The YRCC updates the information of these bankfull discharges every year in its yearly flood control and management strategy before the coming of the flood season.

In case of short term changes, the river bed can have big variations of the cross section shape in one flood event. In case of long term evolution, the change of flow regime, caused by changes of engineering conditions and management strategies, has big impact on the bed level, bed terrain and hereby the flow conveying features. One important change is the bankfull discharge. Before 2002, for decades, the lower Yellow River's smallest safe flow rate, the discharge which is thought to be safe for the floodplains, has decreased from more than 5000 m³/s to less than 2000 m³/s (Li, 2009).

On the other hand the bankfull discharge has a big increase, after the Xiaolangdi reservoir was put into operation in 1999. The reservoir reduces the lower Yellow River sedimentation, in general, in two ways: firstly because it captures sediment from its upper stream inflows, slowing down the velocity in the in-pool water body and making a portion of sediment content to deposit to the reservoir bottom; and secondly because its main function is as a controlling structure for the flow and sediment transport, coupling regulation which aims at producing favorite flow rate and its sediment concentration to transport as more as possible the sediment to the sea.

According to the YRCC, it is known that, at present, for the entire lower Yellow River reach (From Taohuayu a place near the city of Zhengzhou, to the river mouth), the smallest bankfull discharge is now thought to be 3800~4000 m³/s (YRCC, 2009; YRCC 2010). Fig. 4.10 shows the present bankfull discharges of the lower Yellow River.

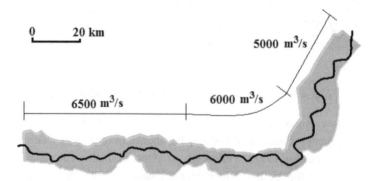

Fig. 4.10. The approximate bankfull discharges of the lower Yellow River (YRCC, 2009)

In the present research, the values of the bankful discharges are (approximately):

(1) 6500 m³/s, bankfull discharge near Huayuankou hydrologic station;

(2) 6000 m³/s, bankfull discharge near Jiahetan hydrologic station;

(3) 5000 m³/s, bankfull discharge near Gaocun hydrologic station.

For the application of the SOBEK 1D2D model, these numbers are regarded as the critical values determining whether the overland flow will be allowed or not. A peak discharge bigger than a bankfull discharge, with enough time steps of equivalent values around the peak, will be used in the inflow hydrograph in order to obtain firm effects of flooding.

In Yellow River the last big flood event was recorded 30 years ago. The last two big floods were Flood 1958 with a peak discharge of 2230 m³/s and Flood 1982 with a peak discharge of 15300 m³/s. Therefore, some flood data used in flood control and management drills are used for the a simple application. Fig. 4.11, 4.12 and 4.13 are three of those hydrographs (i), (ii) and (iii), with peak return periods of below 5 years, 10 years and 50 years respectively under the condition of "natural flow" (before reservoir regulation of the floods).

The model set-up uses for the initial flow, the first value of the upper boundary inflow hydrograph, 4000 m³/s, a value smaller than the smallest bankfull discharge of the lower Yellow River. In a very simple calibration based on the bankfull discharges given above, the bottom levels and upper levels of the 1D cross sections play an important role in controlling the behaviors of the model. After this, hydrographs with different peak discharges are used as upper boundary inflow to run the simulation. Each hydrograph is given enough time steps, namely more than five, of discharges near the peaks. This is to avoid the very sharp peak sections that would make the overland flow hardly happen in especially the condition that a peak is roughly the bankfull discharge at certain river section as shown in Fig. 4.10. The simulation results are shown with Fig. 4.14 to 4.21.

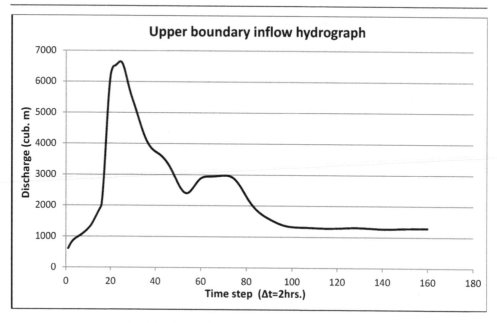

Fig. 4.11. Upper boundary inflow hydrograph (i)

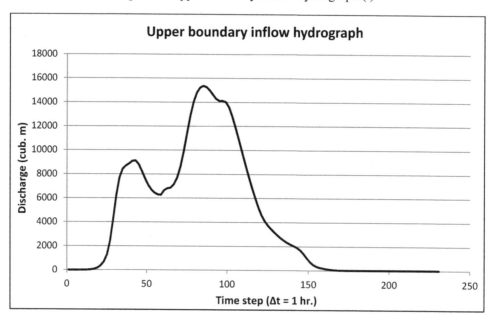

Fig. 4.12. Upper boundary inflow hydrograph (ii)

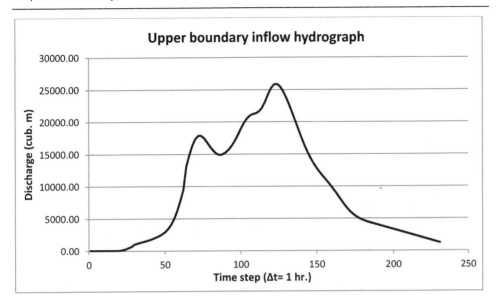

Fig. 4.13. Upper boundary inflow hydrograph (iii)

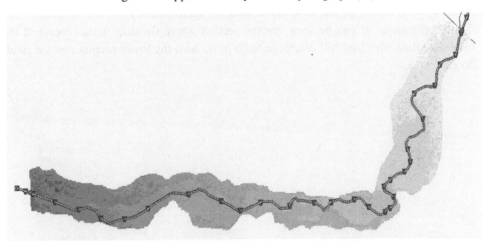

Fig. 4.14. Flooding map of the LYR in case of a peak discharge of 5000 m³/s

In Fig. 4.14, the simulation shows quite good result for the inflow hydrograph with a peak of 5000 m³/s which is the smallest bankfull discharge of the simulation domain. Based on map Fig. 4.10, there should not be much flooding in floodplains and this is consistent with what is shown in Fig. 4.14.

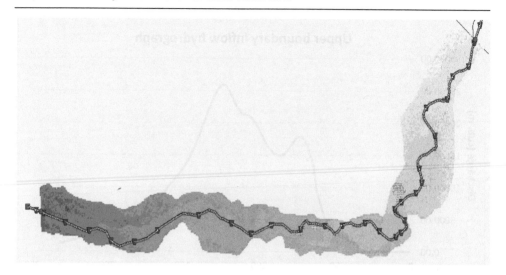

Fig. 4.15. Flooding map of the LYR in case of a peak discharge of 6000 m³/s

In Fig. 4.15 the simulation shows acceptable result for the inflow hydrograph with a peak of 6000 m³/s which is the bankfull discharge approximately at the middle of the simulation domain. It can be seen that the section where flooding occurs locate at the section marked with bankfull discharge 6000 m³/s. Also the lower section can see small flooding area.

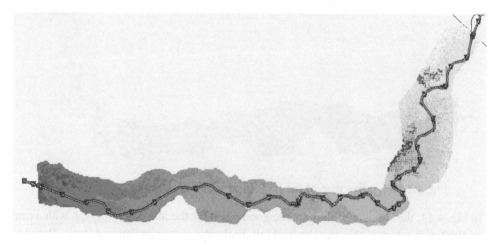

Fig. 4.16. Flooding map of the LYR in case of a peak discharge of 6500 m³/s

In Fig. 4.16, the overland flow happens in a larger scale in the lower part of the domain if compared with Fig. 4.15. This is consistent with the fact that this section has a bankfull discharge that is 1500 m³/s smaller than the peak discharge.

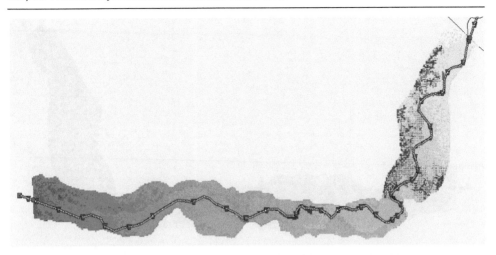

Fig. 4.17. Flooding map of the LYR in case of a peak discharge of 7500 m³/s

In Fig. 4.17, the overland flow happens again much in the lower part. But the middle and upper sections can also see small flooding. This is consistent with the bankfull discharge of 6500 m³/s in the upper section marked in Fig. 4. 10.

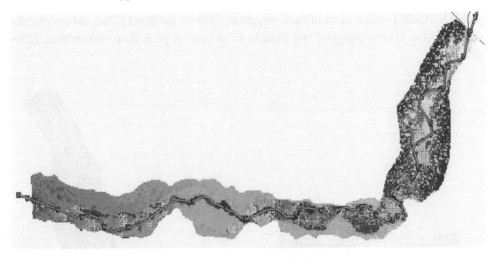

Fig. 4.18. Flooding map of the LYR in case of a peak discharge of 10000 m³/s

In Fig. 4.18, the flooding happens throughout the simulation domain. The upper boundary inflow is much greater than the biggest bankfull discharge 6500 m³/s. There is no experiential number of bankfull discharge, or reliable event-based map for comparison for this case.

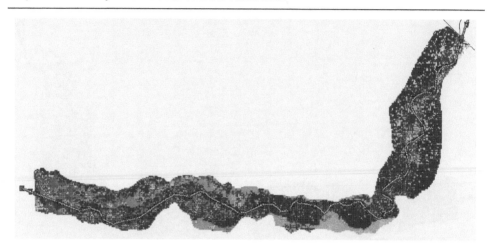

Fig. 4.19. Flooding map of the LYR in case of a peak discharge of 22000 m³/s

In Fig. 4.19, the upper boundary inflow has a peak discharge of 22000 m³/s which is very near to 22300 m³/s, the peak discharge of the flood in August 1958. Most of the areas between the dykes of both sides of the river are flooded, with exception of few high elevation places. There is no doubt that all floodplains will be in great danger and most probably flooded with a flood in such magnitude. This is the flood of the dyke protection standard - the dyke is designed and thought to be safe for peak flow smaller than 22000 m³/s.

Fig. 4.20. Flooding map of the LYR in case of a peak discharge of 25000 m³/s s

Fig. 4.20 has similar situation with Fig. 4.19 with slightly visible bigger flooded area. This flood is bigger than the dyke safety standard, 22000 m³/s. So there is no guarantee of the dyke safety with this situation.

4.5 The 2D flood model using flexible grids

4.5.1 Model setup of D-Flow FM

The uniformed square or rectangular cells schematize the entire simulation domain in the one cell size and shape. But the flow conditions in terms of velocity, and water depth may distribute quite unevenly in a model domain, with some places of rather uniform and mild changes while others of big gradient. For example, in river bends, the flow direction changes greatly while in rather straight and uniform width river sections, the flow velocity changes more gently. A recent major development of 2D modeling is the use of unstructured grid cells. The flexible grids can help schematize the simulation domain with more focus on information of concerned locations. It can also allow the reduction of the runtime of simulation as it can have smaller number of total grids.

A new simulation tool developed by Deltares with flexible grids is now available (Deltares, 2012). This tool can guarantee the combination of different grids so as to properly represent the computational domain and solve the de Saint Venant equations properly. Based on the map obtained for the square cell model, the D-Flow FM was built and run in the same simulation domain. With the application of new modeling tool, new characteristics of the spatial flooding process are defined and presented showing the capabilities of the software application tool in modeling such a complex environment. Due to limited accuracy of DEM and lack of relevant data for the big river, no particular emphasis was made to the accuracy of the model. But the application principle has been shown clearly.

The D-FLOW FM model was built with a numerical schematization based on a Finite Volume Method (FVM) solver of the vertical integrated 1D-2D shallow water equations (Kramer and Stelling, 2008; Kernkamp et al, 2011). An unstructured grid developer was integrated inside the application which enables to perform conversion of coordinates between geographical and projected coordinates.

Both the SOBEK and D-Flow FM 2D modeling tools share the same numerical engine. The major difference is the discretization method, namel, the way the grids are represented and developed. One of the advantages in the use of unstructured grids relies in that narrow areas can be easily refined leaving coarseness in areas where no detail is required (Kernkamp et. al., 2011).

In the simulation domain, no dyke failure is considered. The is based on the fact that the present Yellow River dykes were built to defense a flood of once in a thousand years under the regulation of the reservoir group. This flood is already a quite rare flood. Aiming at serving real-time operation, this research deals with floods which are supposed to be smaller than those with that frequency. Therefore, the dykes are the domain boundaries. No water would go to outside of dykes on both sides of the river. Or in another word, flood propagation develops only inside the area between dykes.

Two major steps were made to obtain the unstructured grids. The first step is to curvilinear grids in the main channel. This was done based in a combination of SRTM data and YRCC survey data. The second is to obtain the triangular girds for the lowland areas. These two types of grids were developed in such way that one type of them keep raw and the other fine, at the same time, keeping alignment and boundary domain the same but increasing the number of grid cells for the later. A selection was made between different grids by means of a measure called orthogonality, which is defined as the consine between the lines that joins two circumcenters of cells and the link between the two nodes that makes them adjacent.

In general, the measure should be zero as in the case of square grids. However, for convention, it is acceptable for it to be less than or equal to 0.05. The measure is selected in such way that a reduction in the uncertainty associated with the computational diffusion of mass and momentum in the numerical solution of the model could be achieved.

In the end, the raw unstructured grid was selected to be used to perform the flood simulations due to the low increase in orthogonality obtained, and to represent meandering areas of the Yellow River (Fig. 4.21).

LOCATION OF NODES IN GRIDS

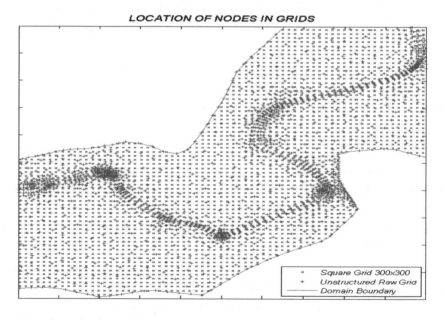

Fig. 4.21. Discretisation grids near Jiahetan station, (x) square grids and (+) unstructured raw grids

4.5.2 Simulation results

As mentioned above, the inundation prediction modeling alone is a big topic for the Yellow River with complex heavy sediment loaded flow and morphological conditions. In the D-Flow application, the emphasis is to show it is one of the tools for operation and to show the application principle. Therefore, the research is not seeking an accurate operational solution but rather to show the general principle of its function and application in management and decisions. Based on this, different hydrographs for the upper boundary of the simulation domain are prepared to represent different reservoir operational results.

These hydrographs are divided into six groups according to their patterns. Thus six scenarios are used to run the model. The result maps are obtained to show the general differences of scenarios. The downstream boundary conditions are set based on the rating curve at Gaocun hydrological station, which is obtained from field surveys. Initial state of the river is determined by running initially a steady flow simulation so that only the influence of the incoming hydrograph could drive flooding in the river. The six maps below show the maps of the maximum flood areas.

Scenario 01: fast concentration time, fast flooding of area but minimum spatial coverage for this event (Fig. 4.22). The simulation shows that for this hydrograph 28.43% of the area is flooded.

Scenario 02: slow concentration time, slow flooding of area but with low spatial coverage of the flooding (Fig. 4.23). The simulation shows that for this hydrograph a maximum flooded area of 37.80 % is obtained.

Scenario 03: maximum time of concentration, very slow flooding of the model but medium to large spatial coverage (Fig. 4.24). The simulation shows a maximum flooded area of 71.72 %.

Scenario 04: slow concentration time, slow flooding in the area but with high spatial coverage (Fig. 4.25). The simulation shows a maximum flooded area of 91.12 %.

Scenario 05: fast concentration time, fast flooding of area and high spatial coverage (Fig. 4.26). The simulation shows a maximum flooded area of 88.32 %.

Scenario 06: intermediate concentration time, with intermediate to large spatial coverage (Fig. 4.27). The simulation shows a maximum flooded area of 68.30 %.

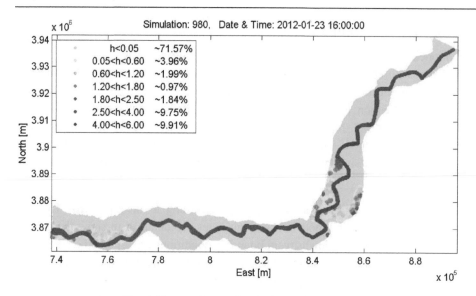

Fig. 4.22. Maximum flooded area for Scenario 01

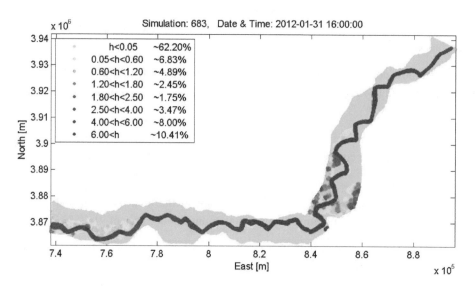

Fig. 4.23. Maximum flooded area for Scenario 02

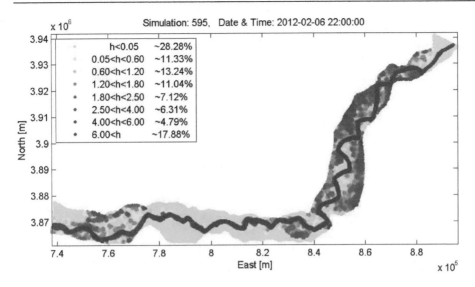

Fig. 4.24. Maximum flooded area for Scenario 03

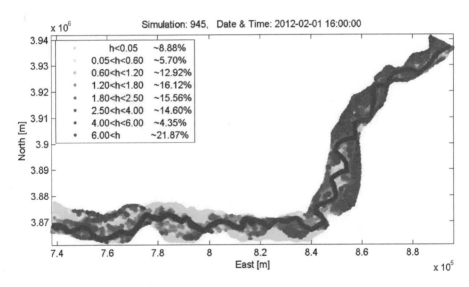

Fig. 4.25. Maximum flooded area for Scenario 04

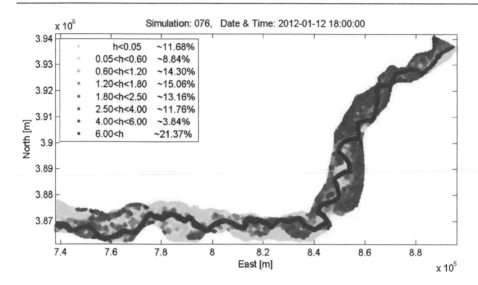

Fig. 4.26. Maximum flooded area for Scenario 05

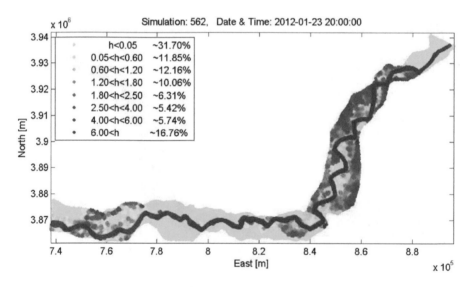

Fig. 4.27. Maximum flooded area for Scenario 06

With the simulation results analysis of the variables related to the flooding events, including the maximum flooded area (FMA), the maximum volume of water inside the domain (FMV), inundation distribution patterns, etc was done. One phenomenon seems to be special was that once the water surpasses the front edges of the floodplains, it would be rapidly conveyed to the areas near the dykes. Then, flows form long the dykes. This is due to the reason that, in our model domain, floodplain terrain is somewhat special - the areas near the dykes are lower than the areas near the main stream. This can be seen with Fig. 4.1. This is caused by the high sediment concentration and sedimentation of the river.

The D-FLOW model used for simulating the flooding on the Yellow River is able to represent different scenarios of flooding inundation in wide long rivers. The freely DEM data downloaded from the SRTM can be used for setting up a pioneer applications. But higher resolution data is needed to build real operational tool. Surveyed data is also necessary for showing the real conditions. After this, operational solutions can be obtained with adequate calibrations.

4.6 Some relevant aspects in real-time operation

(1) Lead time and simulation

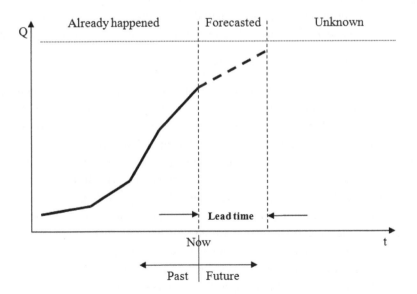

Fig. 4.28. Lead time and forecasted hydrograph in real-time operation
(adapted from Li, S., 2008)

In the presented scenarios of inundations, the used input hydrographs are complete flood time series, as for the cases of Fig. 4. 11, 4.12 and 4.13. In real-time operation, however, the condition would probably be somewhat different: in most cases, the lead time of a formal operational forecast is usually much shorter than time span of a full flood event. Therefore, the forecasted result hydrograph, with much less number of time steps, illustrated in Fig. 4.28 with dashed line, will be given at any moment when it is required by the operators or decision makers.

Each of this forecast starts at the time of "now" and extends to the "future" in a time length of the lead time. Fig. 4.28 shows the idea. After one forecast is given, another forecast some time later may made based on new development of flood. Hence, the forecast will be updated whenever needed as a flood develops. Therefore, at any moment, the model simulation in real-time operation will be made based on given time series, with certain lead time, which is most probably shorter than a complete flood event time span.

(2) The issue of the "farm protection dyke"

In addition to river training works which cannot be reflected on the 300 m X 300m DEM map, the "farm protection dyke" also adds difficulty for fine prediction of floodplain flooding. The "farm protection dyke" is a kind of small and informal dyke which were/are constructed by local farmers at the front (near flow side) edges of the floodplains. They are usually few meters wide and 1-2 meters high. The existence is officially not permitted. However, they are popular in the lower Yellow River, simply because the local farmers can use these small dykes to protect their farmland with certain floods. These dykes make flooding of floodplains a more human-impacted and stochastic.

(3) Macro view effect of the simulations

As has been mentioned, the DEM data is not fine enough for an accurate simulation. A more accurate simulation would require much smaller cell size. In the lower Yellow River channel, for example, there are many river training works in the front edges of floodplains or some key sections near the dykes where the flow direction or dyke safety influenced by flow is key concerns of river safety. These training works usually have a size of around several meters wide, meters high and dozens of meters long. The 300 m X 300 m square cell size, for example, is too big to reflect them, but they are important for the flow's behaviors. At the same time, even if adequate resolution data is available, the simulation computation for such big river section would still be quite difficult for normal single desktop or notebook computer based computation. The real-time operation requires a much shorter time, if compared with general research activities. This research just shows a general solution for the real-time operation, but not a practical operational tool. A feasible operational solution would be putting the model simulation work to high performance severs or some supercomputing machines.

4.7 Summary

Numerical simulations are very useful tools to evaluate the impact of different reservoir simulation alternatives. They should also be a basic component of the management and decision support tool. Nowadays advanced numerical simulation models are available. The prediction of the inundation pattern on the lower Yellow River was done using the state of the art simulation tools produced by Deltares Software Center, in particular SOBEK 1D2D and the new D-Flow Flexible Mesh.

The two models, one with uniform grid and the other with flexible grid, show the general principle of how the alternatives can be evaluated in terms of inundation and flooding process. More work needs to be done to develop real operational tools with much smaller cell sizes and enough calibration, but the feasibility is shown here.

The flexible grids helps focus more on the representation of the flow fields based on different flow conditions and also allows the reduction of the number of cells. Due to limited accuracy of DEM and lack of relevant data for the large river area, no particular emphasis was given to the accuracy of the model. But the application principle has clearly been shown here.

5 A flexible reservoir routing method

Following up from the requirements formulated in chapter 3 for a more flexible reservoir routing scheme analysis, this chapter introduces a new and practical method of reservoir routing, named 'Cross Line Method' (CLM), which is based on solving simultaneously the system of equations for mass conservation and reservoir storage-release. The storage-release relation need not necessarily be defined as an explicit function, as is often assumed in traditional reservoir routing methods. A detailed description of the method is given in this chapter along with calculation examples. An analysis is performed to compare with other methods. The new method allows considerable flexibility to the particular shape of the storage-release relations. And most importantly, it obtains the solution in an analytical way. It is simple and accurate, without bringing more calculation. The calculation process can be easily coded and run on a computer.

5.1 Introduction

The storage-release relation varies depending on the different combinations of available outlet facilities, which may consist of controlled or uncontrolled sluiceways (tunnels) and spillways. The sluiceways may include generator sluices, bottom sluices (e.g. for flushing sediment), irrigation sluices, flood control sluices and so on. A reservoir may have several kinds of release facilities. In practice, since the storage is a bi-jective function, or a one-to-one correspondence between storage level (stage) and volume, the storage-release has a similar relation as the level-release relation. The storage-release relation is used in the routing process, where both storage and release discharge are variables of the control equation and reservoir characteristics. The release is a function of storage level (or storage) if the spillways and the sluices have no gates (i.e. uncontrolled) or if they are with constant gate openings for the entire time that the reservoir is in operation. However, if the release is done through various gate openings operated for different level variation ranges, then the level (storage) release relation becomes very complex. It may contain sections, or zones of different release policies. Here, a section or zone refers to a storage level variation range where the release policy is different above or below a threshold, e.g. constant release discharge below and non-constant releases above.

Different combinations of releases, through bottom outlets or controlled spillways, will produce different level (storage)-release relations. At the same time, in many cases these relations are not easy to be used in general routing solution methods, due to the irregularity of these relations. Li and Mynett (2009a; 2009b; 2010) present the particular requirements. Traditionally, there several methods have been developed that are commonly used in the computation the routing of flow through a reservoir, including the mass-curve method and the storage indication (Puls) method (Puls, 1928). Fenton (1992) shows that the routing can be solved analytically if the reservoir characteristics and inflow can be expressed in functions. However, strictly speaking these methods use the condition that the release is a function of level or storage of the reservoir, which means that any given level or storage within the reservoir will have only one corresponding release discharge. The storage-release 'curve' thus has to be smooth enough and without sharp angles. In practice, for many reservoirs in the world, the conditions are more complex, different controlled release operations and different level zones leading to different release schemes. These schemes are based on operating gates with either uncontrolled or controlled release or a combination of both. If the gates are operated during the routing period, then the routing process becomes complex (Chow, 1964).

5.2 Background

5.2.1 The governing equations for routing calculations

The present chapter starts from the conservation equations introduced in Chapter 3, i.e. equations (3.1) and (3.2). In the routing calculations, the storage-release relation for the particular reservoir characteristics is used. This relation brings the complexity to the calculation.

In equation (3.2), for any time step t, both the release discharge q_{t+1} and reservoir storage volume V_{t+1} at the next time step are unknown. In order to solve the equation the level-release (H-q) relation or storage-level (V-q) relation of the reservoir is used jointly with equation (3.2). Here, the concepts of 'relation' and 'function' are defined as: a relation is an arbitrary set of ordered (x, y) pairs while a function is a set of ordered (x, y) pairs in which each x-element has one and only one y-element associated with it. In traditional routing methods, q is a function of storage volume V (or storage level H). Examples for this function are represented in figure Fig.5.1. However, this assumption is not met in some practical reservoir operation.

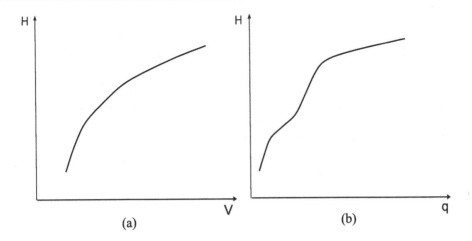

Fig. 5.1. Example: graphs of level-storage and level-release relations

The notation for a relation is V~q relation or V-q relation

or

$V_{t+1} \sim q_{t+1}$ relation (in accordance with equation (3.2))   (5.1)

If q is a function of V, then the notation for it is:

$V = f(q)$ or $V_{t+1} = f(q_{t+1})$ (5.2)

If equation (3.2) is used together with relation (5.1) or function (5.2) we obtain two equations (relations) with two unknowns, which will lead to the solution for q_{t+1}, V_{t+1}. Then the corresponding (H_{t+1}) can be calculated by interpolation from the characteristics describing the reservoir.

The general assumption here is that the reservoir has a horizontal water surface, i.e. the inflow and outflow will not produce big surface slope in the pool, so the water level and storage volume have a good relation. This is a common assumption for most of the reservoirs after the design phase and will not cause large computational errors.

5.2.2 Variables of the reservoir routing equations

There are three kinds of data involved in the reservoir routing process: (i) time series of inflow hydrograph(s); (ii) initial condition data and (iii) reservoir information. The characteristic of a reservoir will include at least two relations: the storage level-storage volume (H~V) relation and the storage level-release (H~q) relation.

The level-storage relation is usually a rather simple and regular curve in which the storage increases up to a maximum value, as the water level in the reservoir increases. The level-release relation generally includes the individual release features of the various release structures and the total release capacity of the reservoir. The release capacity is the maximum release of a reservoir, with all release structures open. An actual release in practice will be a summation of releases of those being opened or operated. The usual form is generally in discretized values, where each level has an associated storage volume and one or more release numbers. In Table 5.1 three types of outlet facilities are presented (q_s, q_f, q_g), and the maximum release is a simple summation of all. It has to be mentioned that Table 5.1 gives only a very simple example whereas large multi-function reservoirs could actually have more release structures and thus many combinations of opened gates.

Table 5.1 Example characteristics of a reservoir

H	V	q_s	q_f	q_g	q
level	Reservoir storage	Spillway discharge	flood sluice	Generator(s) sluice	total release (if all opened)
(m above MSL)	(10^8 m³)	(m³/s)	(m³/s)	(m³/s)	(m³/s)
410	0.253	0	0	90	90
405	0.635	0	140	110	250
420	1.248	0	352	130	482
425	2.183	0	565	140	705
430	3.387	0	698	150	848
435	4.817	0	825	165	990
440	6.698	1980	973	175	3128
445	8.892	2210	1108	190	3508
450	11.330	3490	1193	200	4883
455	12.927	3950	1257	202	5409

The given input for starting the flow routing through a reservoir consists of times series data of the inflow hydrograph. The most common is the forecasted and/or designed inflow hydrograph. In real-time operation of a reservoir, it is often the forecasted inflow hydrograph that is taken into consideration.

The initial water level in the reservoir is given as an initial condition for the start of the routing computations. All the other relevant initial values used in the computation are obtained from this given initial water level in the reservoir. These commonly include the initial storage volume and the instantaneous discharge release from the reservoir.

The variables that are determined through the routing process are the time series of the outflow, and the variation of the storage volume and water level in the reservoir in time.

5.3 The reservoir level-storage-release relations

The assumption that the release is a function of storage volume or storage level is necessary for using a traditional reservoir routing method. For example, in 'Hydrology - Principles, Analysis, Design' (Raghunath, 2006), this condition is clearly stated before the methods are addressed. This approach is crucial to determine whether a reservoir routing method can be used or be used in the right way, before looking into the detail of any operational routing process. In this research, it is found that in some cases a single-valued mathematical relationship between storage volume and release discharge does not exist. This is the main reason for the exploration of a new method.

The level-release relation, unlike the level-storage relation which is a rather simple and regular mono-increase bi-jective function, may have some variation and can be complex if the different combinations of the opening of release structures are made. The release structures might consist of hydropower generator gates, bottom gates for flushing sediment, general-purpose release gates, spillways etc. Under controlled release operation, the release can vary from zero to the release capacity (maximum release at a given level).

For example, the Xiaolangdi reservoir on the Yellow River in China has 3 sediment tunnels, 3 orifice tunnels, 3 free-flow tunnels, 1 spillway and 6 power tunnels. Different combinations of these facilities under uncontrolled release operations will lead to different release schemes. Thus, the actual operational level-release relation, a regulated level-release relation, varies with different combinations of release structures. For some reservoirs, these relationships may have special forms, where the V-q relation can be in a straight and vertical direction or even bend backwards in special sections (see Fig. 5.2 (a), (b), (c) and (d)).

An example is the constant generators' release in a level section. If the total release is formed only by the generators and the generators need less flow power at higher level, then the required release may get smaller as the storage volume become larger. Fig. 5.2 shows examples of the level-release relation where all four graphs represent the resultant relation (summation). In Fig. 5.2 (c), the B-C section is a vertical line where the release remains unchanged when the storage volume increases. This may happen when the number of generators in operation is determined and each of these generators has a constant discharge rate. Fig. 5.2 (d) shows another example in which the release even decreases when the storage rises from point B towards C. The Sanmenxia reservoir, on the Yellow River, will have such behaviour if the total release is only through the running generators, below a certain level.

Note that, in this research, the V-q graphs are used several times to show the conditions and derivations of the method.

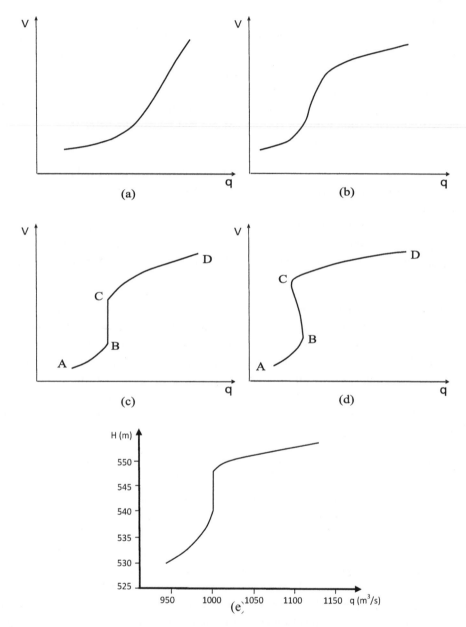

Fig. 5.2. V -q relation examples, (a), (b), (c) and (d) and the particular case of the V-q relation for Guxian reservoir (e)

The existing method based on uncontrolled release can be used for routing with the V-q graphs patterns shown in Fig. 5.2 (a) and (b), but not for those in (c), (d) and (e) where the release is not a function of level (or storage) for the entire characteristics' domain (from point A to point D). An example is the operational H-q relation of Guxian reservoir on the Yiluo river branch of the Yellow River (Fig. 5.2 (e)). In this case, the release from 520 m to 529 m is a smooth curve while it becomes a constant value of 1000 m³/s from 529 m to 548 m.

As a typical example, in real-time operation the reservoir release is usually given in the form of H-q relations rather than as uncontrolled release schemes, in which constant discharge releases or release time series are given. From a computational point of view, the actual uncontrolled release details are not given in this case and they are not necessary for the computation aiming at real-time operation and decision-making. Therefore, it is necessary to develop a method that can deal with controlled release, which may consist of two or even more sections of different types of V-q relations.

It has to be mentioned here is that controlled release can be divided into two different types, i.e. (i) with given H-q relation and (ii) with no fixed H-q relation. For the latter, the only way to do routing is making the computation with particular rules. So, this chapter deals with computations that have a given fixed H-q relation.

5.4 A new routing method: the Cross Line Method (CLM)

Based on the principle of conservation of mass, equation (3.2) can be rewritten as:

$$V_2 = -\frac{\Delta t}{2} q_2 + \left(V_1 + \frac{Q_1 + Q_2}{2} \Delta t - \frac{q_1}{2} \Delta t \right) \tag{5.3}$$

where V_1, Q_1, q_1 are the instantaneous values at the start of the time step,
while V_2, Q_2 and q_2 are the values at the end. For one computational step,

$\left(V_1 + \frac{Q_1 + Q_2}{2} \Delta t - \frac{q_1}{2} \Delta t \right)$ is constant,

because V_1, Q_1, Q_2, q_1 and Δt are all known. Therefore, equation (5.3) is a linear equation for which the general form is:

$$y = -ax + b \tag{5.4}$$

where

$$a = \frac{\Delta t}{2}, \qquad b = V_1 + \frac{Q_1 + Q_2}{2} \Delta t - \frac{q_1}{2} \Delta t \tag{5.5}$$

This line has a slope of -a. Since Δt is positive, the line's direction will always be from upper left to lower right in the coordinate system, see line A-B in Fig. 5.3. The smaller the Δt, the flatter the line will be. At the same time, the V-q (or H-q) graph will extend from near the origin to the far upper-right, because the release discharge will generally increase as the storage volume increases. So, we have the following two equations:

$$\begin{cases} V = -aq + b \\ \quad V = f(q) \end{cases}$$ (5.6)

Or

$$\begin{cases} \qquad V = -\frac{\Delta t}{2}q + b \\ V \sim q \text{ relation of characteristics} \end{cases}$$ (5.7)

where, V and q are V_2 in q_2 in equation (5.3) respectively.

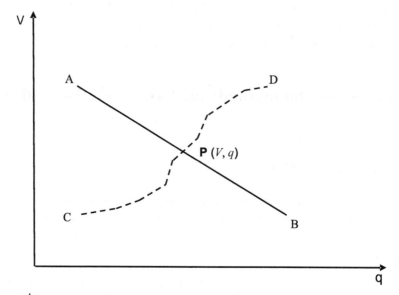

Legends:

$$—\text{ Line A-B: } V = -\frac{\Delta t}{2}q + b, b = \left(V_1 + \frac{Q_1 + Q_2}{2}\Delta t - \frac{q_1}{2}\Delta t \right)$$

$$-\,-\,\,\text{ Polyline (fold line) C-D: } V \sim q \text{ relation of characteristics of}$$
reservoir

Fig. 5.3. Arbitrary reservoir V-q relation and linear mass conservation equation

In practice, the V-q relation is generally given in discrete form, so if we look into the basic segments of the graphs, they are actually (piecewise linear) straight lines. The V-q relation (C-D) is a 'polyline' (folded line), which is composed of segments of straight lines. In Fig. 5.4, the line between P_i and P_{i+1} is a straight line.

If we can find the two points that define this line at the crossing with the mass conservation equation, then we have the equations of two linear functions with two unknown variables V and q. Therefore, we can easily get our solution in an analytical way by solving the two simultaneous equations of the linear relations, or by linear interpolation. This is the basic starting point of the Cross Line Method for reservoir routing developed here.

Once we know the intersection points, we have the coordinates for V and q and hence our solution. Fig. 5.4 shows a more detailed sketch of the method. The V and q relationship, such as the V and q number pairs in Table 5.1, is drawn in a multi-segment form with discrete data. Each segment is a straight line (piece-wise linear function). From the lowest level to the highest, we have point $P_{c,0}$, $P_{c,1}$, ..., $P_{c,n}$. Here the point number starts from 0, simply because we want to take the first point $P_{c,0}$ as a relative origin.

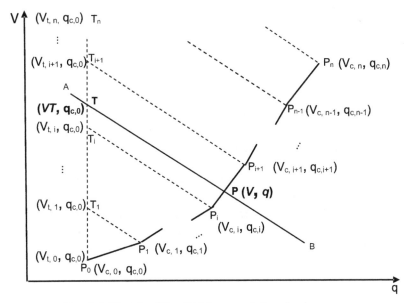

Fig. 5.4. Sketch of the CLM computational scheme

From equations (5.3) and (5.4) we see that all mass conservation equations have the same

$$\text{slope} = -\frac{\Delta t}{2}.$$

101

The only factor that distinguishes one equation (for a time step) to another is b, expressed in equation (5.5). We can make use of it if we draw lines with the slope $-\frac{\Delta t}{2}$ from each V and q relation point (P_0, P_1, ..., P_n) to a vertical line which starts from P_0, (line P_0-T_n).

From basic algebra we know the equation for a line passing through point P_i has the form:

$$\frac{V-V_{c,i}}{q-q_{c,i}} = -\frac{\Delta t}{2} \tag{5.8}$$

or in more general form:

$$V = V_{c,i} + \frac{\Delta t}{2}\left(q_{c,i} - q\right) \tag{5.9}$$

where V and q are unknowns and the others are known variables. Here V and q are V_2 and q_2 in equation (5.3) resp. They are the instantaneous values at the end of the time step being calculated. Equation (5.12) is the equation for lines passing through the characteristics points (P_0, P_1, ..., P_n). It is obvious that these lines are all parallel lines with slope $-\frac{\Delta t}{2}$. Their intersection points with the vertical line P_0-T_n are T_1, T_2, ... Tn.

Next, we calculate the V and q coordinates for T_1, T_2, ... Tn. For the vertical line P_0-T_n, the straight line equation is q = $q_{c,0}$. So, the q values of the coordinates are all known. Then the V values can easily be obtained by substituting q = $q_{c,0}$ into the straight line equation (5.9), which leads to:

$$V_{t,i} = V_{c,i} + \frac{\Delta t}{2}\left(q_{c,i} - q_{c,0}\right) \tag{5.10}$$

After calculation of each line with this equation, we got two sets of coordinate pairs, the storage V and release q coordinates for points P_0, P_1 ... Pn on the folded-line graph P_0-P_n and the storage V and release q coordinates for points P_0, T_1, T_2, ... T_n on the straight line P_0-T_n. These provide the basis for the routing method.

For each time step in the reservoir routing process, we may have a different line segment and intersection point, so the reservoir routing calculation always goes from one time step to the next one. Once the values are obtained, we get equations in the same form for the next time step and so on and so forth, up to the end of the time series. The water level can be calculated as a linear interpolation in the level-storage relation.

5.5 Step by step algorithm of the proposed CLM method

With reference to Fig. 5.4 again, the detailed steps of flow routing in a reservoir, based on the proposed CLM method are:

(1) Calculate V_t and add V_t values to the reservoir's characteristics

Substitute each release value of the storage-release relation into equation (5.10) - the V_t values are obtained. Put them into a new column in the reservoir's characteristic table (see e.g. Table 5.2)

(2) Calculate the 'locator' value V_T for the time step being calculate

Let $q=q_{c,0}$ (i.e. $q_2=q_{c,0}$) and substitute into the mass conservation equation; we get:

$$VT = -\frac{\Delta t}{2} q_{c,0} + \left(V_1 + \frac{Q_1+Q_2}{2} \Delta t - \frac{\Delta t}{2} q_1 \right) \tag{5.11}$$

This VT locates the cross point of line A-B and line P_0-T_n. VT is the value only for the current time step being calculated, and different steps would have different value of VTs. So it is calculated in each time step.

(3) Find the V-q characteristic's index numbers (e.g. i and i+1) between which the cross point locates

This is done by comparing the VT value and the V_t values from $V_{t,0}$ to $V_{t,n}$ (see Fig. 5.4 and Table 5.2). If VT is found between two adjacent values making as $V_{t,i}$ and $V_{t,i+1}$, then we know that the line cross point is located between these two points. Here the values at time step i (and therefore also at i+1) are kept for later calculation use.

(4) Calculate unknown value V and q for the present time step

In this step, the location of the straight line of the mass conservation equation has been found precisely by the locator VT. Based on this, the values for V and q can be obtained in different ways, which fundamentally share the same mathematical background. Two different ways of calculation are given below.

In order to present the scheme clearly, before giving the calculation, Fig. 5.4 is simplified to be a more calculation-oriented sketch as shown in Fig. 5.5. Since line T_{i+1}-P_{i+1} is parallel to line T_i-P_i, line A-B must cut line sections $T_{i+1} - T_i$ and $P_{i+1} - P_i$ in the same ratio. Then the calculation schemes are:

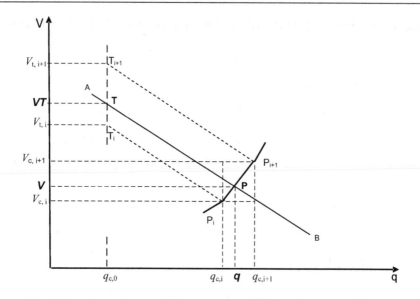

Fig. 5.5. Computational scheme with linear interpolation (V, q are unknowns to be found)

(a) Solve by linear interpolation

From Fig. 5.5, we have:

$$\frac{VT - V_{t,i}}{V_{t,i+1} - V_{t,i}} = \frac{q - q_{c,i}}{q_{c,i+1} - q_{c,i}} \tag{5.12}$$

or:

$$q = q_{c,i} + \frac{VT - V_{t,i}}{V_{t,i+1} - V_{t,i}} \left(q_{c,i+1} - q_{c,i} \right) \tag{5.13}$$

Here q is the solution for the release discharge. Since $V_{t,i+1} - V_{t,i} \neq 0$ throughout the computational domain, this equation is always valid for every section of the V-q curve, even with vertical and backwards bending sections.

Now, the value V can be easily obtained in the following way:

- By substituting q into formula (5.10)

$$V = V_1 + \frac{Q_1 + Q_2}{2} \Delta t - \frac{q_1 + q}{2} \Delta t \tag{5.14}$$

or

$$V = -\frac{\Delta t}{2} q + \left(V_1 + \frac{Q_1 + Q_2}{2} \Delta t - \frac{\Delta t}{2} q_1 \right)$$

- By linear interpolation with the storage volume

$$\frac{VT-V_{t,i}}{V_{t,i+1}-V_{t,i}} = \frac{V-V_{c,i}}{V_{c,i+1}-V_{c,i}} \tag{5.15}$$

which gives:

$$V = V_{c,i} + \frac{VT-V_{t,i}}{V_{t,i+1}-V_{t,i}}\left(V_{c,i+1} - V_{c,i}\right) \tag{5.16}$$

- By linear interpolation method with release q

$$\frac{V-V_{c,i}}{V_{c,i+1}-V_{c,i}} = \frac{q-q_{c,i}}{q_{c,i+1}-q_{c,i}} \tag{5.17}$$

which becomes:

$$V = V_{c,i} + \frac{q-q_{c,i}}{q_{c,i+1}-q_{c,i}}\left(V_{c,i+1} - V_{c,i}\right) \tag{5.18}$$

Though equation (5.18) is an option for calculation, it is not recommended for general use. The reason is that $q_{c,i+1} - q_{c,i}$ could be zero, as for example, when the V-q curve becomes vertical (see section B-C in Fig. 5.2 (c)). In this case, the release remains unchanged over a certain range of storage variation. This is a problem with some old methods of reservoir routing. Therefore, the last option is not suggested.

(b) Solve the problem with an analytical solution of two straight line equations

In this way, the mass conservation equation, which is a straight line equation with two unknowns V and q, is put together with another straight line equation of P_i and P_{i+1}. For the latter, since both ends of the line have known coordinates, $(V_{c,i+1}, q_{c,i+1})$ and $(V_{c,i}, q_{c,i})$, the equation can be written as:

$$V = V_{c,i} + \frac{q-q_{c,i}}{q_{c,i+1}-q_{c,i}}\left(V_{c,i+1} - V_{c,i}\right) \tag{5.19}$$

This is exactly the same expression as equation (5.18); but here q is still unknown. Now, we have two variables and two equations:

$$\begin{cases} V = V_1 + \frac{Q_1+Q_2}{2}\Delta t - \frac{q_1+q}{2}\Delta t \\[2mm] V = V_{c,i} + \frac{q-q_{c,i}}{q_{c,i+1}-q_{c,i}}\left(V_{c,i+1} - V_{c,i}\right) \end{cases} \tag{5.20}$$

By transforming the mass conservation equation by using b expressed in equation (5.5) and then solving equations (5.20), q is obtained first and then substituted back into the mass conservation equation to get V.

In this way, the answers for q and V are:

$$
\begin{cases}
q = \dfrac{(b-V_{c,i})(q_{c,i+1}-q_{c,i})+(V_{c,i+1}-V_{c,i})q_{c,i}}{\frac{\Delta t}{2}(q_{c,i+1}-q_{c,i})+V_{c,i+1}-V_{c,i}} \\[3mm]
V = V_1 + \dfrac{Q_1+Q_2}{2}\Delta t - \dfrac{q_1+q}{2}\Delta t
\end{cases}
\qquad (5.21)
$$

where, b is calculated from equation (5.5).

This is the analytical solution of the Cross Line Method. Here, 'analytical' means that we solve the problem in a piece-wise discrete form. If the data is accurate enough, our solution here should be accurate enough as well. The values V and q can be directly calculated when the time step is known. Here V and q represent the instantaneous values at the end of the time step, usually marked as V_2 and q_2 or V_{j+1} and q_{j+1}. The index j is the particular time step. Once V and q (i.e. V_2 and q_2 or V_{j+1} and q_{j+1}) are found, they can be taken as the known instantaneous values for the start of the next time step, the V_1 and q_1, or V_j and q_j and so on.

5.6 Discussion

5.6.1 The flexibility of the Cross Line Method

The new method was developed to have more flexibility than the existing ones. The Cross Line Method obtains the result in an analytical way by finding out the point of intersection of the mass conservation equation and the V-q graph (line) of the reservoir characteristics. Therefore, it allows for routing processes with multi-sectional V-q graphs where each section of the graphs can have different attributes of the V-q relations. If the mass conservation equation line has a cross point with the V-q relation, the solution will be obtained irrespective of the types of the V-q relations. So, even if the outlet is not a function of water level (or storage), i.e. controlled release, the method can still work without precondition. The only condition is that the release has 'some' defined relation with storage or storage level.

In Fig. 5.6, the V-q relation graph has 4 segments, A-B, B-C, C-D and D-E. Each of the sections represents a release scheme of the reservoir by a combination of its release facilities in a either uncontrolled or controlled manner. For the entire valid level variation extent (from A to E), the release is in controlled mode. A section's character may reflect some operational information, for example, for a certain outlet facility (Fig. 5.6), the D-E section might imply the operation of spillways, because in this section the total release increases sharply as the storage becomes larger.

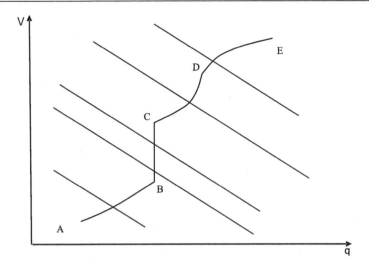

Fig. 5.6. Cross Line Method with multi-sectional V-q relation

In all 4 segments, B-C is supposed to be vertical - the release does not change (grow) when the level (storage) gets larger. The intersection points of the mass conservation equations with the characteristics graph (here, perhaps 'graph' is better name than 'curve') are our solutions. It can be seen that it works perfectly with different V-q relations, including the vertical one. In fact, it can also work for the special situations as shown in Fig. 5.2 (d) if the time step is chosen in such a way that the slope of the mass conservation equation line is bigger than the section's slope. The reason for this is that it makes the straight line of mass conservation equation more horizontal than the characteristics graph section so that they are sure to have a cross point. This condition is:

$$-\frac{\Delta t}{2} > S_g \tag{5.22}$$

Where, Sg is the slope of the characteristics line section. If the V-q relation is given in discrete form as in most of the reservoirs, S_g of a section can be easily calculated with two adjacent characteristics data. Take the example in Fig. 5.4, the slope S_g for the graph (here is straight line) between Pi and Pi+1 is:

$$S_g = \frac{V_{c,i+1}-V_{c,i}}{q_{c,i+1}-q_{c,i}} \tag{5.23}$$

in which $q_{c,i+1} - q_{c,i} \neq 0$ (if $q_{c,i+1} - q_{c,i} \neq 0$ is the vertical case).

In the Cross Line Method, the release q can be obtained by two methods, either by equation (5.13) or by the q expression in equations (5.21). In the first method, the only denominator is Vc, i+1 - Vc, i, which is always bigger than zero. So, there is no possibility that a division by zero happens.

For the second method, the denominator is:

$$\frac{\Delta t}{2}\left(q_{c,i+1} - q_{c,i}\right) + V_{c,i+1} - V_{c,i} \tag{5.24}$$

Vc, i+1 - Vc, i is always a positive number, so if Δt/2 (q_(c,i+1)-q_(c,i)) is negative, it is possible for expression (5.24) to give zero. However, that happens only when the following expression holds:

$$\frac{\Delta t}{2}\left(q_{c,i+1} - q_{c,i}\right) = -(V_{c,i+1} - V_{c,i}) \tag{5.25}$$

After transformation, this becomes:

$$\frac{V_{c,i+1} - V_{c,i}}{q_{c,i+1} - q_{,i}} = -\frac{\Delta t}{2} \tag{5.26}$$

This is actually the slope of the straight line equation of mass conservation. With expression (5.22), we already make sure that this will not happen. Then, if the second method is used to obtain q under the V-q relation shown in Fig. 5.2 (d), a check is necessary to make sure that Δt is small enough so that the lines will have a cross point.

Fig. 5.7 shows the effect of using time interval Δt in different sizes. Line A-C via B is the V-q relation of reservoir's characteristics. Fig. 5.7 (a) indicates that a smaller Δt gives a more horizontal line for the mass conservation equation, while a bigger Δt makes a more steep one (in order to show the slope difference, the three lines all pass the same point).

In Fig. 5.7 (b) and (c), the mass conservation lines are consecutive lines of computational time steps. From the graphs, it can be seen that smaller Δt means more horizontal mass conservation line and more delicate routing process while bigger Δt implies more steep mass conservation line and coarser routing process. Therefore, in general, a smaller Δt would make a more accurate calculation (but it should not be too small to avoid significant numerical calculation errors).

It has to be mentioned here that smaller Δt may require resampling of the inflow time series. If the original inflow time series is regarded to be accurate enough or unchangeable, then a simple linear interpolation is the general way forward. In general, the time interval of inflow time series is given and a smaller one does mean more accurate inflows. It helps lower the routing errors with the special conditions.

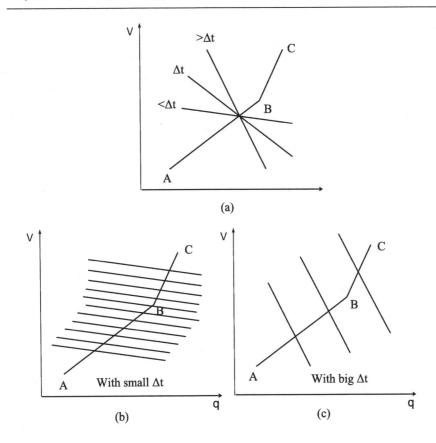

Fig. 5.7. Routing with different time interval Δt

5.6.2 CLM application for special reservoir storage conditions

In general, the reservoir routing process can take the time interval of the inflow hydrograph as its time step. However, in extreme cases, the inflow may increase or decrease very sharply and the characteristics may change rapidly. In that case, a smaller Δt is suggested to make a more accurate routing calculation. These extreme cases only occur with very special conditions, such as very small reservoir storage capacity with very sharp and large flood inflow discharge.

Fig. 5.8 shows the comparison of an ordinary case (a) and an extreme one (b). In both of them, L_j and L_{j+1} are mass conservation lines and their cross points, P_j and P_{j+1}, represent solutions of two adjacent time steps where j and j+1 refer to time step number. In Fig. 5.8 (a), the two lines are rather close due to smaller time interval. Also the neighbouring characteristics sections change their directions from one to another in a mild way. In this case, the release discharge solutions, q_j and q_{j+1} are accurate, because the change from Pj to Pj+1 in a linearized shortcut way (dotted straight line from Pj to Pj+1) is not a big difference compared with the change from Pj to B and then from B to Pj+1. In Fig. 5.8 (b), however, the shortcut (dotted straight line from Pj to Pj+1) is a big difference from starting at P j going to B and then, after a turn, to Pj+1. This can produce a kind of 'pseudo' solution, that is, it looks 'correct', but in fact, the instantaneous values of storage is calculated bigger than it should be and the release is made smaller. It can be seen that, with smaller time step length with which the calculation is obviously more accurate, the release will change more closely along the line from Pj to B, and then B to Pj+1 and thus the reservoir will release more than the calculated result.

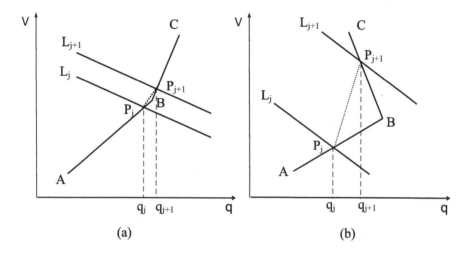

Fig. 5.8. Comparison between general case and extreme case of calculation in terms of different Δt and discretized characteristics data

This extreme case reveals how the size of the time stepping influences the accuracy of the reservoir routing process. It is just an example of those that might be relevant to the use of all reservoir routing methods, not only the Cross Line Method. In general, the change of storage volume resulting from inflow and outflow in a time interval is much smaller than the discrete storage volumes of the reservoir, and therefore the change from P_j to P_{j+1} will not be large. So, the case shown in Fig. 5.8 (b) is a very extreme case and is rarely seen in engineering practice.

In any case if the time interval is chosen properly and the reservoir characteristics are discretized accurately enough, reservoir routing using the Cross Line Method leads to good solutions.

5.7 Conclusions

Reservoir routing is the basic process of calculating the passage of a runoff hydrograph through a reservoir. Traditional methods usually assume that the release is a function of water level or storage. These methods work fine at given unchanged *uncontrolled* release conditions (here uncontrolled means that a certain opening of release facilities, should be maintained – if it is changed, the methods do not apply). A controlled release is even more beyond these methods. However, a real-time operational release scheme may be given, for an operational variation extent, as a combination of some release facilities of a reservoir in *controlled* release, and the result of the combination is no longer a function of the level or storage.

Therefore a new method is required (YRCC, 2007; Li et al, 2012a; 2012b) to deal with a 'multi-sectional' controlled release with given storage-release relation so that the routing process is correct and accurate. The Cross Line Method given in this chapter is exactly such a method. If the controlled release does not have a given storage-release relation, for example a rule-based release based on upstream and downstream conditions, then the routing process has to be made with the actual rules.

The Cross Line Method obtains the storage volume and release discharge by locating the intersection point of the V-q straight line equation of mass conservation and the V-q relation curve of reservoir's characteristics. In the V-q Cartesian plane, the intersection point coordinates V and q are the solution for that time step. The storage level can then be interpolated with the characteristics by using V. So, based on discrete reservoir characteristics data, this method seeks the solution in an analytical way. This method has at least the following advantages over other methods:

- The traditionally required condition that the release is a single-valued function of level (or storage) has been extended to 'some mathematical relation' (not necessarily a single-valued function);
- It is able to deal with a controlled release scheme; the multi-sectional H-q (or V-q) relation with each section allows both uncontrolled or controlled release operations;
- It is a simple procedure with clear steps in the routing calculations;
- It is quite accurate, especially if the calculation is done in an analytical way;
- It directly deals with V, q variables rather than with indirect functions of them;
- It can deal with special V-q relations such as a constant or even decreasing q with increasing storage;
- It does not have the potential danger of division by zero (robust procedure);

- It does not require more complex calculation;
- It can be done easily with table-based hand calculation (for verification);
- It can be easily coded and processed by a computer;
- It can be represented in semi-graphical way;
- It enables an automatic process of complex release schemes without special arrangement near interfaces between zones with different release policies.

Δt is a factor that may influence the accuracy of all routing methods, including the Cross Line Method. In most cases when the change of volume in a time interval is much smaller than the reservoir's instantaneous storage volume at that moment, there is no need for employing a smaller Δt. However, if the change is large over the storage volume, errors could be significant due to the size of the discrete step; in that case, a smaller time step is suggested. The Cross Line Method allows a variable time step and is a perfect routing method in such cases.

6 Next generation software architecture for decision support

Numerous flood control and management software systems have been developed over the past decades to support the decision making process, in particular for large river basin authorities. However, many of these systems were found not to be useful in real practice, in particular under tense conditions, e.g. in crises situations. A rapid and robust DSS tool is needed in those conditions. Based on practical experience in the Yellow River and other large river systems in China, this chapter explores some of the shortcomings of the present-day decision support systems for flood management and introduces a new approach to increase their applicability. A real-time reactive mechanism is proposed. Two simple applications are presented to demonstrate the approach and some preliminary conclusions are drawn on how to develop next generation decision support systems by improving user interactions based on this new mechanism.

6.1 Introduction

6.1.1 Adaptive management application

In this research, the consequences of adaptive management are explored for the DSS system architecture and user interaction. The consequences of proposed decisions are computed with the embedded numerical simulation tools as described in chapter 4. In adaptive management, the evaluation of alternatives with their consequences for flood control and reservoir operation determines the management actions.

There have been many studies on how adaptive management can be successfully and correctly applied. According to Walters (2007), adaptive management must include a formalized process of learning, preferably combined with deliberate experimentation. By formalized learning he refers to a comparison with original expectation in order to revise models and/or management actions based on what was learned, including specific articulation of what information is being sought and precisely how it will change future decisions (Runge, 2011).

At the core of adaptive management is a critical assessment of uncertainty that impedes the consequences of a preferred alternative. When decisions are recurrent, implementation coupled with monitoring can resolve uncertainty, and allow future decisions to reflect that learning (Runge, 2011). According to Gray (2005), an important part of an adaptive approach implies that decision makers should ask the right questions. In many situations, natural asset managers may find it helpful to examine requirements related to the spatial and temporal context, corporate culture and functions, data and information management needs, dynamic policy formulation, and effective communication.

6.1.2 Yellow River approach

In a river basin with serious flood threats, like the Yellow River in China, forecasting, control and risk management of the flood are important issues for river basin authorities. For flood forecasting in uncontrolled basins, one needs meteorological predictions, rainfall-runoff models, and a flood routing scheme. However, for flood control and management in regulated river basins, a more advanced system is needed that can explore the effect of e.g. alternative strategies for reservoir control on flood risk mitigation. River systems with flood control structures like reservoirs and gates constitute a complex network for storing, regulating and conveying flows. In such systems, proper operation of the control structures is essential for protecting people and safeguarding relevant regions. In particular as flood risks develops, efficient analysis methods and decision making procedures are of utmost importance for determining the reservoirs' storage capacities and river flow-conveying capabilities (Li and Mynett, 2008).

A number of flood control and management software systems have been developed over the past decades, in order to support the discussion, analysis and decision making, which is now quite popular in large river authorities like the Yellow River Conservancy Commission (YRCC) in China. But, as has been mentioned, to develop a successful DSS is still a big challenge for not only China's river authorities but those in other countries as well. As an example, Uran and Janssen(2003) studied the use of spatial DSSs, including five examples on coastal zone management in Europe and then concluded that some of these systems have not proven to be useful in real practice, in particular in emergency situations, according to end-users and decision makers. Hence many river authorities are working hard to develop new and better systems. In China, from around the year 2000, the implementation of China's National Flood Control and Drought Relief Command System (NFCDRH, 2007) program covers it's seven largest river (lake) systems. One of the key goals is to build a suite of seven flood control and management software systems for each of the seven management authorities. Key questions to achieve this goal are: (i) what are the generic processes in flood control and management practice; (ii) what are the typical problems which prevent some of the present systems from being perceived as useful by the end-users; (iii) what would be the most efficient way to support flood control and management operation; (iv) what kind of software systems are needed for real-time operation; (v) what kind of software architecture and operational features are required to build such system? In order to answer these questions, it is very important to analyze the entire decision making process. Flood DSS systems are built for a rather special group of end-users: from technical specialists running alternative scenario's, to final decision makers at the top level of information-flow. This requires a particular software development approach for flood control and management, as analyzed in this chapter.

6.2 Functional decomposition of a DSS system for multi-reservoir-based regulation

6.2.1 Next generation software environment for river applications

Before any decision process or operational style is carried out, the basic computational functions have to be studied in order to make a good base for the higher level application. A robust computational core lays good foundation for the high level flexible functional composition.

When studying the overall software environment for the Yellow River management, Mynett and de Vriend (2005) suggested that a distributed architecture for networking, processing, storing and integrating data and tools must be built to support the decision making process(see Fig. 6.1). The development of an adequate software environment is absolutely crucial to the success of an integrated river basin model. It is necessary, but by no means sufficient, to develop a number of modules, e.g. describing the flow, the water quality, the sediment transport, the morphology, the ecology, etc. To enable an integrated modeling approach, special attention needs to be paid to the architecture of the software environment and the usage of the component models. In fact, one can discern four layers in the software environment: (i) support layer, (ii) functionality layer, (iii) application layer and (iv) usage layer (see diagram below) (Mynett and de Vriend, 2005).

The 'bottom level', viz. the support layer, closely connects to the conceptual and architectural aspects of the ICT infrastructure (the so called 'backbone'), which is addressed separately below. The "'op level, viz. the usage layer, must support collaboration and workflow management, enabling smart integration of applications in a secure web-based service environment (software portal), and offer intuitive, easy-to-use interfaces through which the users can communicate and add value by contributing.

The Support Layer provides facilities to enable the coupling of the relevant modules, such as: (i) a data communication facility (data transfer between modules, connection with measured data); (ii) auxiliary software, such as input and output facilities (e.g. GIS, output processing facilities, presentation facilities), and necessary middle ware; (iii) a module coupling facility (which information is transferred in what form, when and from where to where); (iv) a knowledge management facility, enabling the gathering, validation, storage and access to knowledge and expertise. The use of modern distributed and networked environments and collaboration between different users and suppliers of data calls for a suitable security strategy on the technical level which may have organizational and institutional consequences on the level of the support layer.

Usage Layer						
function: to facilitate operational use (N.B.: socio-technical aspects)						
result: practical use, feedback from users						
Application Layer						
action: building a model of a specific situation						
result: e.g. Lower Yellow River Model						
Functionality Layer						
Module FLOW	Module WAVE	Module QUAL	Module MOR	Module ECO	Module ...	Module ...
Support Layer ('backbone')						
module coupling and steering facility						
auxiliary software						
communication facility						

Fig. 6.1. Layered software architecture (Mynett & De Vriend, 2005)

The Functionality Layer includes the various modules that describe the physical and ecological processes to be incorporated in the model. A key property of the envisaged architecture is that each of these modules can be replaced by another, if necessary via a wrapper that transforms the input and output and connects it with the support layer. In this way, knowledge and expertise from other parties can be incorporated if necessary.

The Application Layer concerns the actual construction of a model for a specific situation, e.g. the Lower Yellow River or a part of it. It gives access to facilities to select modules, to define their coupling, to generate computational grids, to connect input datasets, to schematize the geometry and to set the model parameters, etc. This will usually be done via a user interface for model construction.

The Usage Layer facilitates a user-friendly environment for the model that has been built in the Application Layer. In the river engineering and management profession, modeling usually concerns the lower three layers. If a model is used in a design study, e.g. in the functional design phase of engineering works, there is a limited amount of time available and the number of runs needed is not very large. In such cases, the Usage Layer may not be felt absolutely necessary, although it may be of use, e.g. in complex, multi-actor decision making processes. This is different when the model is used for operational support purposes, as in the case of Flood Early Warning Systems or hour-to-hour flood management, when the system is used more or less continuously and by many different users. In such a situation, the Usage Layer is indispensable, since it provides the necessary requirements for a multi-user environment. The Usage Layer also provides user-friendly access to the knowledge management system, for interrogating it, or for supplying it with new information.

One important aspect of the Usage Layer is that the issues are not purely technological; social aspects of model usage must also be considered, so that users can derive optimal benefit from the model, give feedback on its fitness-for-use and add new information. In this way, the usefulness of a model can be greatly enhanced. These issues closely connect to those discussed below on implementation. Software relevant to this layer can partly be acquired, but parts that are specific to water management will need to be custom-made. Integrating the designed functionalities and application layers in working environments that fit relevant user groups is a major concern.

Making next generation ICT work involves substantially redesigning water management and business practices, and thus profoundly affecting the way user organizations operate. Next generation ICT can succeed only if infrastructures, data and tools are integrated in a collaborative, workflow-based environment. Such an environment takes users through a number of well-designed steps/phases of implementation, enabling users to "grow" into the new functionalities and suppliers to customize solutions to their customers' needs. This forms the major challenge of the design of a business/service model within an ICT-based environment, called an Application Service Provider (ASP).

6.2.2 The functional structure and basic components

The software environment proposed by Mynett and de Vriend (2005) is at a macro level scope description, being relevant for applications in all relevant fields of river management, with the multi-reservoir-based flood management application being one of its components. Our system would be a particular tool for flood management and decision. It can have similar structure of functional composition, while with more focus on the decision support.

If different tools are studied, it is easy to find that different systems, or applications, require different architectures. Model based simulation systems are discussed more than a decision system is. As an example, Delft-FEWS of Deltares is a software tool that emphasizes its architecture. The philosophy of Delft-FEWS is to provide an open shell system for managing the forecasting process. The structure of the DELFT-FEWS places the data process in the centre rather than puts it around the models as in the traditional model-centered systems (Werner et al, 2004). Various utilities are available to deal with generic processing of data in the context of flood forecasting that interface with the database, as well as an open interface to modeling systems that effectively allows incorporation of a wide range of forecasting models, independent of supplier. Here the data-centered approach is advancement over the traditional model-centered implementation, for it allows more focus on the capability of the open shell.

In this research, the multi-reservoir-based flood control and management tool should be designed with a good mechanism and structure for functional composition, which will support the high level application for management and decision.

The requirements have been discussed in the first two chapters of this thesis and they are the main driver elements for this chapter. A DSS oriented multi-reservoir regulation application means not only the water mass conservation computation of reservoirs and simulation of river flow, but also implies the design and implementation of the complete computational process comprising reservoir routing, individual reservoir rule and control, multi-reservoir collaborative rules and control and reservoir-river system coupling calculation, based on well understanding of the multi-reservoir operational rules, a good simulation model of the downstream flooding extent and software engineering implementation skills. In order to construct a robust, flexible and adaptive functional structure, the computational components will be designed in different logical layers corresponding to different levels of computation and control rules. These components will be made in such way that the lower ones can be linked by control rules to form upper layer ones. An advantage of this is that, various end-user regulation functional requirements can be built by using the lower layer "building block". The linkages among the blocks and components are the management rules. Both the rules and program control of the application form the overall functional structure of the flexible and DSS oriented multi-reservoir regulation computational core.

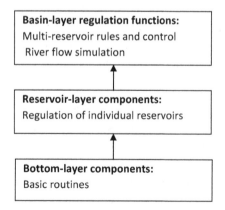

Fig. 6.2. Three layers of regulation computation

The multi-reservoir regulation calculation is made with three logical layers (see Fig 6.2.) They are the bottom-level components, reservoir-layer components and basin-layer functions. The bottom layer components consist of the basic routines for reservoir regulation calculation. Each of them can be easily invoked by the upper layer components, namely the reservoir-layer components.

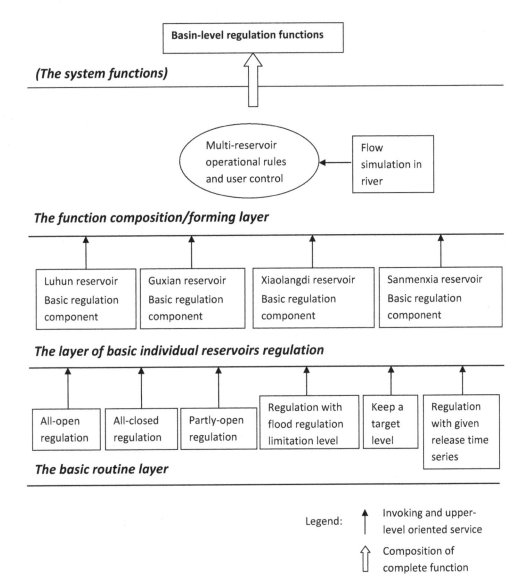

Fig. 6.3. Structure of regulation functions

119

The reservoir-layer components are formed by fundamental regulation calculation of a reservoir. Each of the components is a rather stand-alone program that fulfills a unique function formed and controlled by the reservoir operational rules that are not specified in the basin-layer management rules. The top layer components are the functions that directly serve the users. These components are more complex programs, which are made from the lower layer components with the control rules dominating the computational flow control. Or in another word, these components are organized by the rules. A more detailed sketch is shown in Fig. 6.3.

In the three-layer structure, the top layer, the basin-layer regulation functions, deals with multi-reservoir (basin-layer) operational rules. These rules control and link the lower layer computational components and, at the same time, make check the control targets according to intermediate calculation results to assign more iteration-based calculation till the rules are satisfied. This layer manages the basic computational components, which are lower layer ones. The only computational aspect in this layer is the routing/simulation of river flow in the channels which link the reservoirs and convey flow to downstream.

The computational components in the two lower layers are designed in such a way that the top level layer can always have a complete function to meet the user's need by managing and organizing the lower layer components with operational rules corresponding to that user's requirement. Meanwhile, these computational components are designed to have clear functional definitions and information exchange interfaces. They are designed to be rather independent and standard routines (in C# methods) with clear functional definitions and usage indications. This makes the usage of them a high level application of rule-based computation deployment rather than typical lower level programming.

This design makes the system flexible and adaptable to both rule-based operation and real-time decision support. For the rule-based regulation, this is similar to the HEC-ResSim "user-scripted" rules editing, and DHI Mike Basin tool's "Visual Basic macros" for users to use the rules. But here, we emphasize not only the rule-based regulation or simulation but also the support of decision making process."

6.2.3 River flow routing due to multi-reservoir regulation

Traditionally, there have been many flow forecasting and routing methods developed for operational use. Szilágyi and Szöllösi-Nagy (2010) listed some of the most popular methods based on constant wave speed assumption. These methods include Muskingum-Cunge method, Puls method, Muskingum cascade, 1D and 2D simulation model, etc. In this research, a flow routing scheme used by YRCC is taken as a basic method for the river flow routing. The method is based on Muskingum cascade approach. Among other, Rui (2004) presented the computational details. At the same time, the design of the computational procedure makes it very easy to take other method as well. The reason for using the scheme is that the scheme is within the protocol between flood forecast and flood management in YRCC. If we take the flood forecasting as input of the management system, a river flow routing method in accordance with the protocol is a good choice for the consistency.

The schematization of the multi-reservoir river system is shown in Fig. 6.4 where each rectangle represents a river reach, i.e. there will be one river research for the confluence coefficient method of flow routing. The parameters were obtained by calibration with historical floods. The usage of the method and parameters can provide a good basis for the coherence of river routing calculation with the real-time operation. This method can be replaced with other flow simulation tools if they are well calibrated and proved good enough for practical use.

It is important to mention here that the resultant hydrograph at Huayuankou is calculated in an add-up summation way, though there are other methods to do it. The reason for this is that, in this way, the formation of the hydrograph can be easily studied and analyzed.

Fig. 6.4 is, as well, a schematization based on the internal protocol between hydrological forecasting and flood control management. The obligations defined in this protocol are: (i) the hydrological department (i.e. the Hydrological Bureau of YRCC) provides the flow forecasting for the sub-catchments of Sanmenxia to Huayuankou reach, numbered from Q1 to Q11, including the flow from upper main stream of the river, namely the inflow of Sanmenxia reservoir, numbered Q10. So, they are the ten discharge time series of rainfall runoff flow forecast and one flow forecast from upper main stream; (ii) simple summation method is used for the calculation of the time series at joint points. Then the resultant new time series can be routed to further downstream. The triangles in Fig. 6.4 are these summation calculations; (iii) the flow forecasting lead time refers to that of the Huayuankou hydrologic station which is the key reference station for the operation of resesvoirs. In YRCC, the standard lead time for mid-lower Yellow River flood management is decided to be 8 hours according to the river network and flood wave propagation. But for early warnings, longer lead time is used with less accuracy.

The numbering of all time series is shown in Fig. 6. 4. This numbering will be used throughout this research and thesis when referring to a discharge time series. Table 6.1 presents the parameters of the flood routing method for Sanmenxia to Huayuankou flow routing computation.

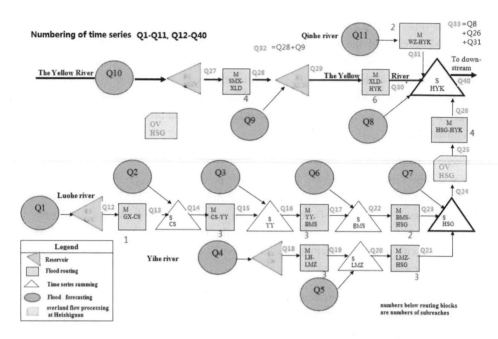

Fig. 6.4. Computational flow schematization of the Sanmenxia to Huayuankou multi-reservoir river system

Table 6.1 The cascade flood routing parameters

River reach	Number of sub-sub-reaches	K	X	delta t
Sanmenxia_Xiaolangdi (SMX-XLD)	4	1.9	-0.1	2
Xiaolangdi_Huayuankou (XLD-HYK)	6	2.3	-0.1	2
Luhun_Longmenzhen (LH-LMZ)	3	1.8	0.4	2
Longmenzhen_Heishiguan (LMZ-HSG)	3	2.2	0.35	2
Guxian_Changshui (GX-CC)	1	2	0.5	2
Changshui_Yiyang (CS-YY)	3	2	0.35	2
Yiyang_Baimasi (YY-BMS)	3	1.5	0.3	2
Baimasi_Heishiguan (BMS-HSG)	2	2.6	0.33	2
Heishiguan_Huayuankou (HSG-HYK)	4	2.3	0.3	2
Wuzhi_Huayuankou (WZ-HYK)	2	2.3	0.3	2

6.3 Decision making process in flood control and management

Decision making processes within different river authorities may vary from one another in terms of style and content. Still, usually three stages are distinguished in the decision making: (i) pre-flood preparation, (ii) operational flood management and (iii) post-flood assessment (Dahm, 2006). When developing a practical DSS for, one has to carefully study all underlying processes and relevant aspects. The following analysis is based mainly on the practice in flood control and management at the Yellow River Conservancy Commission and similar large river authorities in China.

Fig. 6.5 shows the flow chart of a typical process of flood control and management process for river authorities in China. Before any decision is made, collecting information and flood forecasting are the main activities.

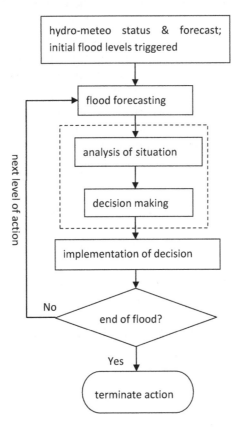

Fig. 6.5. Typical flood control and management process of river authorities in China

The information usually includes real-time meteorological data and forecasts, real-time hydrological information, real-time and predicted status of control structures, real-time status of damages, losses and resilience actions, etc. Specific site information may also be collected, if needed.

Flood forecasts usually consider different lead times, ranging from short-term formal forecasts, to early warnings as well as mid-term predictions that are all key information components for decision making. In some river authorities, the various forecast functions are distinctly separated. Usually, the formal forecasts are required to meet strict accuracy standards and should be issued by specialist forecasters. Such formal forecasting plays an essential role in the subsequent decision making process.

Having obtained all necessary information, the decision makers have to assess the main issues as well as the control and management objectives for the specific situation. Key issues comprise proposed changes in reservoir operation and identification of potential detention areas (Shu and Deng, 2000). Based on the overall objectives and available measures, the decision makers and experts will develop several possible alternative measures corresponding to different management considerations. These alternatives are then evaluated and compared. For large river systems, the alternatives may involve technically complex computational processes, e.g. reservoir regulation calculations, flood simulations, GIS-based hazard prediction, etc. Although sometimes decision makers may have a general idea of how flood control structures could be operated, they may not be perfectly clear about the detailed operational implications in terms of targeted storage level, discharge, inundation area etc (Fig. 6.6). Here computer based management systems with embedded simulation and calculation models can help to provide more clear information.

Decision making is the process of attending to the facts and details, developing practical and realistic alternatives, while minimizing risks and losses. If a decision is made collectively, the decision makers will discuss and analyze possible choices jointly and share views with each other before making a final decision. The decision making process will follow the real-time development of a flood in an iterative way. In general, decision making will continue when there are new/updated forecasts. The time between two consecutive decision making actions is not a fixed interval. It will be determined by the actual development of a flood and the opinions of decision makers as well.

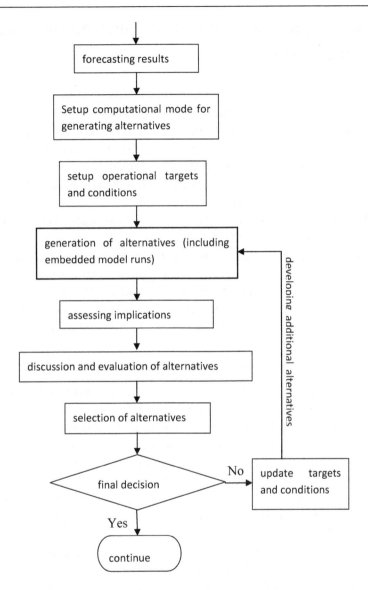

Fig. 6.6. Key stages in the flood control and management decision making process

6.4 User requirements on a DSS for flood control and management

Despite the popularity of DSS software systems in numerous application areas, there are many decision support systems that are never or hardly used (Uran and Janssen, 2003). As an example, over the past decade there has been a rapid development in the field of Integrated Spatial Decision Support Systems (ISDSSs). Although they are gaining attention in the planning and policy making community, there are few examples of their actual use (Van Delden, 2009). A critical assessment of the literature related to the area of decision support systems reveals that the designs of many such systems focus on the available technology rather than on the type of support required by the decision maker (Mann et al., 1986; Keen, 1997; Sammon, 2001). Decision support systems may be somewhat different from other software applications for they are much easier to be rejected by end-users, but the prevailing question is: what are the reasons behind the failures of DSSs? If one wants to build a successful system, the problems preventing it from being useful have to be known. A closer look is presented hereafter on the problems encountered with DSS systems aimed to support flood control and management decisions.

6.4.1 User requirements in terms of operational details

Operational decision support for flood control and management has to focus on a special group of end-users, both technical specialists and final decision makers, which are sometimes referred to as "top level users". For the design and implementation of any DSS there should be direct involvement of these target users. Any successful operational system cannot be developed without involving the actual end-users. In fact, it has been shown in numerous cases that regular interactions and communications between software system developers and target users are absolutely necessary to keep any development on the right track. Decision makers in flood control and management are a particular group of special end-users: they are often not interested themselves in dealing with certain technical issues in the design and development phase, but end to focus more on practical operational details. That is, for decision makers in flood management, the DSS is an application that others prepare for them. Decision makers may have some general ideas and background knowledge on the DSS, but not much in technical detail. This leaves major challenge when designing a flood control and risk management DSS. Developers cannot build a system by simply guessing the needs of the end-user. Therefore, detailed user requirement specifications on content, performance and display style are key to build a successful DSS.

6.4.2 System architecture and execution procedure

Decision support systems generally consist of four major components: (1) user interfaces, (2) databases, (3) modelling and analysis tools, (4) the DSS architecture and connectivity network (cf. Sprague and Carlson, 1982; Power 2007). This traditional statement already reveals the importance of the system architecture. The first three components are basic elements connected by the system architecture, which usually receives less attention and study. But, it is one of the key factors determining the success of a DSS.

From examining a number of systems supporting decision making processes in the fields of flood control and management, it was found that researchers and software developers often build a decision support system in much the same way and same style as numerical modelling and simulation systems (Moya Quiroga et al, 2013; Popescu et al, 2012) . A typical example of such system architecture is illustrated in Fig. 6.7.

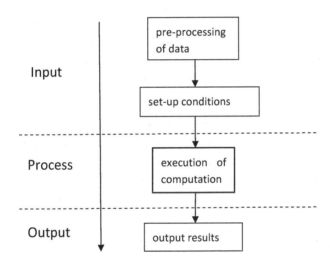

Fig. 6.7. A simple example of a system architecture following IPO (Input-Process-Output) sequence

In this case, the simulation model is at the center of the process while the system architecture follows the 'batch-like' execution procedure of the model. This can be referred to as a 'simulation-centered' architecture. These types of systems are very popular for model-based applications such as hydrological or hydrodynamic modeling systems. They follow the traditional procedural programming approach well known to most researchers. However, this paradigm may not be the ideal architecture for decision support systems, which not only contain simulation models but also have to meet specific requirements (Jonoski and Popescu, 2011) for more direct man-machine interaction and interactive system response. Investigations of existing DSSs show that the IPO architecture leads to a long sequence of processes which is not favorable for optimizing and improving alternatives in an efficient and interactive way. Hence a new approach with different system architecture needs to be explored.

6.4.3 User interfaces and graphical display

A user interface is another basic component of a DSS. Unlike a conventional numerical simulation system, a DSS for flood control and management is not solely a simulation tool, it also should accommodate the process of generating alternatives, evaluating scenarios, analyzing flood control measures and supporting the decision making process along the way. Hence, DSS interfaces for flood control and management should be designed and implemented to deal effectively with very high level information display rather than low level data. However, it was found that many DSSs for flood control were equipped with user interfaces which are similar to a numerical simulation system, and not suitable for decision making, since they usually (i) contain too much information; (ii) require too lengthy operations; (iii) are lacking good graphics. All these are not needed for the decision making process, especially in formal collective decision meetings at large river basin authorities.

No one would argue that the user interfaces of software applications have been changing considerably over time in terms of style, performance, interaction etc. With the advances in cyber hardware and processor speed, the software applications have changed as well towards a direction that is more intuitive and user-friendly. A typical example is the Apple iPhone which became popular to the world in just few years. Yet, in flood control DSSs it is nowadays still very common to see a user struggling to learn the usage of a particular software system. In fact, many users dislike the way they are asked to operate such systems, on which developers may have spent considerable amount of resources. Obviously DSSs for flood control and management also require much better user interfaces and graphical displays.

6.4.4 Use of IT resources and tools

For flood control and management DSSs, the use of IT resources and tools has to be carefully considered. Only those that really contribute to the usefulness and efficiency are to be employed in a DSS. A case study carried out on one of the DSS systems in the Yellow River showed that misuse of network-based distributed databases led to considerable extra work, which would not have been necessary if the system had required only limited amount of data. The particular system used a data sever which was linked to the intranet to serve the DSS as a software client. But it required considerable effort to maintain the server, which employed the large database system under an operating system that was not familiar and frequently hit by a virus from the net. Moreover, further examination showed that neither the database system nor the net-oriented server hardware were necessary. An alternative and simplified solution based on local storage of relatively little data proved to perform even more effectively!

The use of advanced network technologies should serve the real needs for better decision making rather than become a fashionable display of latest technologies. Flood control and management are the responsibility of specialized departments of river basin and reservoir authorities. In these departments, the users of the system are often limited to the actual decision makers and their expert staff during joint decision making meetings. Such system does not have to be 'fully-network-based', but can be designed in such a way that only limited external data is imported via the network, while scenario development and graphical display can be addressed locally. In this way the performance and stability will increase tremendously.

6.5 A new approach in developing better DSS

Based on examining the problems with existing systems and analyzing the generic processes of decision making, a new system architecture and corresponding user interface style as well as operational mechanism are proposed here as an effort to improve the practicability of flood control and management support systems. A simple example of a test program for single-reservoir operation was developed to demonstrate the concept. The test program was evaluated by a group of experts at the YRCC who supplied suggestions for further improvement. The general comments were that 'considerable progress in practicability was achieved' and that the suggested approach would provide an excellent basis for developing a multi-reservoir flood control and management system.

6.5.1 The real-time reactive mechanism for DSS

Fig. 6.4 shows a sketch of the real-time reactive concept that was developed. In general, computerized systems can be divided into three broad categories: (i) Transformational systems that compute output from inputs and then stop (e.g. numerical computations and compilers); (ii) Interactive systems that interact with their environments at a pace dictated by the computer; and (iii) Reactive systems that continuously react to stimuli coming from their environment by sending back responses (Berry, 1999). Such reactive systems are purely input-driven and must react at a pace dictated by the environment. A real-time reactive system continuously reacts to stimuli received from the environment by sending messages, where each reaction must be made within certain time bound (Wan, 2002). Real-time, reactive approach seems a most appealing basis to create a really advanced and yet practical DSS system for flood control and management, in view of its correspondence to the generic nature of the underlying decision making process.

Particularly in case of a collective decision making process, all relevant aspects of flood management are examined and discussed step by step. As a solution is dawning, more and more specific details are taken into account. Finally, the focus will be on the discussion of only a few issues that remain to be important. Decision makers may have become clear about the operational aspect of a particular alternative, but may not be able to provide the operational results in terms of discharge time series, target water levels or storage level changes at each river section, control point or reservoir. Here the embedded simulation models in a DSS come into play. The real-time reactive mechanism illustrated in Fig. 6.8 can be employed for supporting decision makers who then can focus on deciding on alternatives.

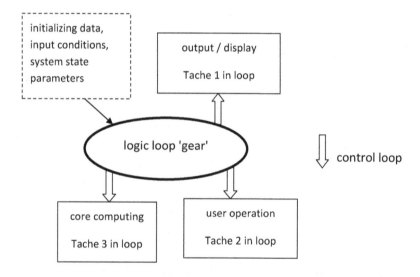

Fig. 6.8. Real-time reactive mechanism for DSS system architecture

After initializing the system with available data and parameter settings, the system starts to run a loop, in which the core computing represents models and simulation components, which usually require the most computing capacities (but which can be reduced by using database technologies or model reduction techniques or artificial neural network emulation to optimize the process); the user operation component represents the continuous input of decision variables that the users want to examine in order to improve any alternative; the output / display presents the resultant behavior of the alternative. Once in the loop, the operation of the system and its response is in real-time reactive mode, facilitating the support for decision making in a super interactive and efficient way.

6.5.2 End-user based graphical interfaces for decision making

A user interface is the "window-to-the-world" of a software system to serve the users. User interfaces largely determine whether a DSS is useful or not. Bad user interfaces can easily spoil the entire system. There is not a fixed standard for user interfaces of DSSs. Different decision support systems have quite different design of user interfaces. Typical Windows interfaces with menus and buttons on top usually lack temporal and operational sequence information and do not properly consider the special requirements of flood control and management DSSs. Therefore, a different approach is necessary to meet the decision makers' requirements. The approach chosen here is to follow the practical tradition of graph-based operation. In this case all users, both decision makers and technical experts, can obtain their information quickly in a familiar way on computer displays and large-screen projections in the decision room. This requires developer to study the use of graphs in the user's tradition rather than taking a common programming tool to make a general graph.

The operational procedure for a single reservoir is shown in Fig. 6.9. The program fulfills the two design features: (i) a real-time reactive system architecture, and (ii) an end-user requirement based graphical interface (see Fig. 6.10 for a screen snapshot).

The reservoir is considered to have the function of regulating floods. For given inflow forecast, the decision makers have to decide on the release pattern (discharge time series) of the reservoir. First, the program will work out a solution based on given conditions and the reservoir's general operating rules. This solution can be regarded as an initial alternative. Then the decision maker would evaluate it and propose changes based on calculated time series for reservoir routing results downstream. In Fig. 6.10, the blue line represents the inflow forecasting; the green line is the storage level for that alternative; the red line is the release discharge of the alternative. The circle and the thin lines crossing it are the cursor position indicator. Beneath the graphs, the numbers in red, blue and green correspond to the release discharge (interval average), release volume (interval average) and inflow forecasting (discharge).

The numbers and their input domains are not meant for inputting the time series manually, but to monitor and show simultaneously (in real-time) the values of the cursor position. However, manual input is available with these domains. When the cursor is left-pressed, the release discharge number will immediately be repositioned to the cursor location. At the same time, the calculation of the reservoir routing will be executed to obtain the new resultant storage level, and all the graphs and numbers will be updated to the new values in real-time. The reservoir routing calculation could be seen as a model simulation. In the past, a simulation-centered approach was usually followed with clear steps of preparing the input data, running the simulation model and displays the output results. But in this example, the simulation model is put into an event-driven loop.

131

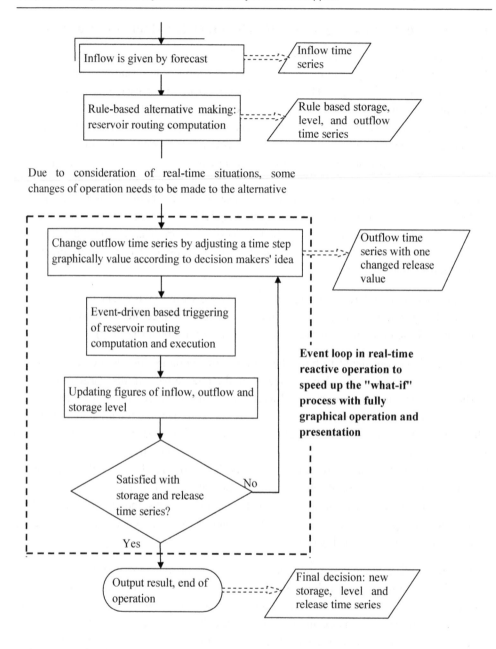

Fig. 6.9. A simple example of single reservoir operation decision with real-time reactive mechanism

One can hardly notice the simulation process since it is in some sense "hidden" in the loop and executed in such a short time that the end-user can hardly notice it. This is a fundamental change in system execution mode. In this way, the decision maker can quickly change the release pattern, and obtain an update of the results in a real-time reactive way. Therefore, he would be able to optimize the alternative to his personal preference and experience. Such approach corresponds exactly to the objective that a DSS supports the decision making process rather than making decisions. The sample application was perceived to be very interactive and much appreciated by the test group at the Flood Control Office of YRCC, who recommended this approach to be introduced.

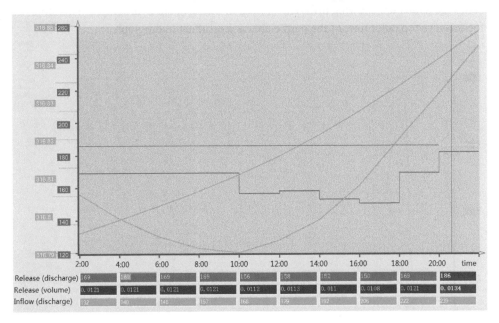

Fig. 6.10. Graphical user interface for the real-time reactive operation of a single reservoir

In case of multi-reservoir operation, the decision makers need to decide the release pattern (discharge time series) of several reservoirs for a real-time flood event. In this multi-reservoir operational condition, however, they have to study the entire river-reservoir system before any decision can be made. The operational scheme of each reservoir is the key issue in the decision making process. In most cases, as this process goes on, more and more facts are made clear, for example, the 11 forecasted local inflows from sub-basins (including inflows to the four reservoirs), which parts of the catchment will get bigger floods and which parts not, etc. Then, the necessary operations that do not have disagreements will be discussed first, for example, a reservoir with small inflow will have less concern and also decision of operational scheme would be relatively easier. After that, some key issues will be studied based on all relevant information. In this example, the operational scheme of Xiaolangdi reservoir will be exemplified, after decisions of the other three reservoirs in the system, are made.

133

Fig. 6.11 shows three source (contributing) hydrographs (excluding release from Xiaolangdi) for Huayuankou hydrograph. Fig. 6.12 shows the summation of the three contributing hydrographs at the Huayuankou station where a complete hydrograph would also need to add another hydrograph from Xiaolangdi's release, after routing in the Xiaolangdi to Huayuankou river section. (The map has been shown in previous section of this chapter for the combination of the Huayuankou hydrograph.) In Fig. 6.12, areas in different colors are time series which forms the Huayuankou hydrograph. These time series are added up, showing different sources of the resultant hydrograph. Therefore, the top edge, or top outer line, of the purple area is a resultant hydrograph.

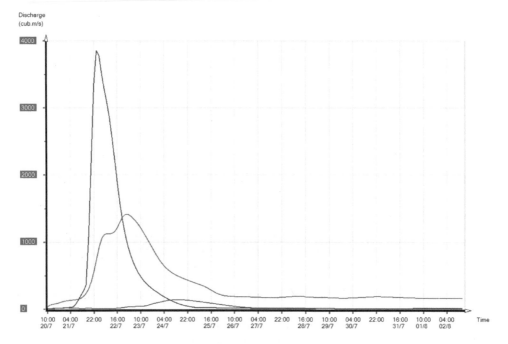

Fig. 6.11. The three discharge sources

Now, suppose the decision makers wants to know how big and what pattern the Xiaolangdi's release, especially the part that will contribute to the resultant hydrograph near its peak section, should be in order to have a more satisfactory resultant Huayuankou hydrograph. Here, it is also assumed that the operational scheme of other three reservoirs has been decided. So, at this moment, the main concern of the decision making process is to determine the release pattern of the Xiaolangdi reservoir. More specifically, the main task is to optimize interactively the release discharges of Xiaolangdi reservoir which will contribute to the hydrograph at especially the peak section of Huayuankou hydrologic station.

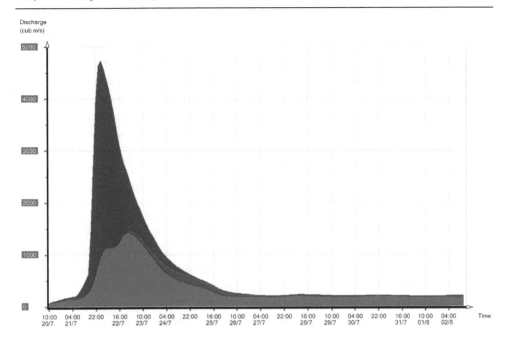

Fig. 6.12. The Huayuankou resultant hydrograph as addition of the three sources

The general operational rules for Xiaolangdi reservoir will be the basic guideline for the decision making while reasonable changes of operation can be made based on consideration of real-time situation. With the general operational rules during flood regulation time, the making of the operational scenario should be focused on: (i) to reduce the peak section of the Huayuankou hydrograph, because bigger peak section discharge may imply bigger threat of inundation floodplains if the peak and its nearby flow exceeds lower Yellow River bankfull discharge; (ii) to discharge Xiaolangdi's flood storage when the forecast is given that the discharge of Huayuankou will definitely decrease due to prediction indicating quick and definite shrinking of upcoming precipitation and hydrologic scales, because the rules require the flood volume detained in reservoirs during (flood-season) regulation operation should be discharged as early as possible so as to make the reservoirs ready for future possible floods. Or, in another word, the "flood regulation pool" is required to be recovered as soon as a flood process is supposed to an end based on formal reliable forecast. One common requirement is that the discharge of this storage should not create another "artificial" flood peak.

(ii) Decision support and alternative development with real-time reactive mechanism

Traditional ways of supporting the above-mentioned decision process would be: (I) first, give a value, run the system and check the result. If not satisfied, another round of operation is carried out, and so on; (ii) make the process programmed with some typical

135

operational scenarios, run the system and see the solutions. If not satisfied, make another version of scenarios and another run; (iii) work out solutions automatically by software tools and let computer give the best answers. These methods may look quite well from general simulation or research points of view, however, in a real flood decision making process, especially the collective decision making process, they may not response quickly enough with the decision making that requires high efficiency of interaction between man and machine. Also the answers might be unacceptable to the decision makers. A few discrete, limited "what-if" simulations and their results might not meet requirement of the real-time decision making process, which involves many ideas and arbitrary "what-if" execution requests in every aspect of the flood in a very short response time.

Therefore, the approach here is to provide an utmost response speed for the support of the decision making process. The process will not be in the IPO pattern, but rather in the real-time reactive mechanism. A quick, handy change of value of a variable would be responded with no time delay in real-time reactive mode. This is sure to be better than any discrete "what-if" solutions, because it can cover any value and it is in a mode of real-time instant result showing on demand. In this way, the displaying of the result is linked closely to the change of the input. Any change of the release discharge, the controlling variable, would lead to immediate change of the resultant hydrograph. In the mean time, the resultant graphs are updated in very short time and the user can hardly feel the delay. This example shows a process that has much complex model computation. It consists of multi-reservoir routing calculation, river system flood routing and time series summing, etc. It can be seen that the models, such as flow routing in river channels and regulation calculation with reservoirs, are all included in this process. The model calculations are made less visible in this process and they are no longer the center of the operation - they are just the necessary computation hidden in the process. The focus here is the quick response of the system to its users' request.

The Microsoft Windows double-screen was used in order to have a good view and convenient control of variables (discharge values), see Fig. 6.13 (a) and (b). One window (display) is used for controlling the Xiaolangdi's release while the other one is responsible for showing the resultant Huayuankou hydrograph in real-time reactive mode. This double-screen presentation mode can be easily implemented with the Microsoft Windows operating system on a PC, if we have one extra monitor. The use of double-screen proves to be a perfect solution for the real-time reactive mechanism operation, which requires both manipulation of operation and presentation of resultant information at the same time. Also, more information can be displayed with the added area of a second screen. This is a simple and practical solution to the real-time reactive DSS approach if with enough computational capacity support.

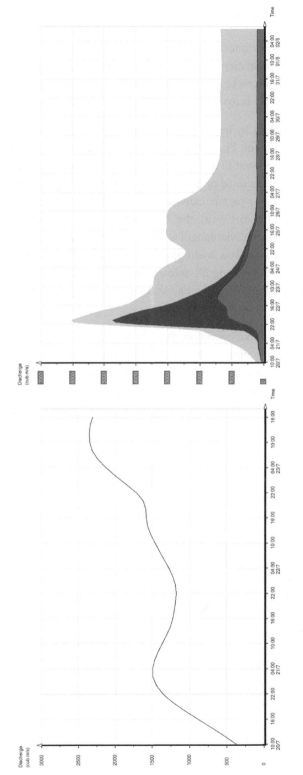

(a) View in the controlling display. The blue line is Xiaolangdi's release (shown only first 40 time steps of the entire time series) and it is the control factor that can be changed to optimize the Huayuankou hydrograph.

(b) The resultant Huayuankou hydrograph shown in real-time reactive mode in a second monitor. The blue area represents the flow from Xiaolangdi's release (after the Xiaolangdi to Huayuankou flow routing.) The lower three layers are the same as in Fig. 6.11.

Fig. 6.13. The double-screen Windows displaying application for real-time reactive mechanism of decision supporting

137

Although the controlling windows interfaces and the hydrograph displaying can be put into one monitor as our most applications are, the use of double-screen can substantially increase the displaying area and make the operation and displaying more fluent and comfortable without bringing more programming work and computational capacity.

Fig. 6.13 (b) shows the complete hydrograph, which are formed by four contributing discharge time series. Compared with Fig. 6.11, one more is added and it is the flow from Xiaolangdi's release, i.e. the green part in the graph. In the decision making process, when only the Xiaolangdi's release is examined or discussed, the shape of this part would be the main focus. The "behavior" of this part is caused by the manipulation for Xiaolangdi's release. Change of blue line in Fig. 6.13. (a), the Xiaolangdi's release, means the change of the upper boundary inflow of the Xiaolangdi to Huayuankou river section. Then, the contributing hydrograph to Huayuankou hydrograph would also change after the river routing process. The graph of Xiaolangdi's release should be changed easily with a mouse. The feeling of the operation would be like that the user designs or draws a graph with a given drawing tool.

With the support of real-time reactive operation, the Xiaolangdi's release can be changed and optimized with a direct observation of the resultant hydrograph at Huayuankou. The shape of hydrograph can be modified with the operation of the release interactively without delay. Fig 6.14 shows another resultant hydrograph at Huayuankou.

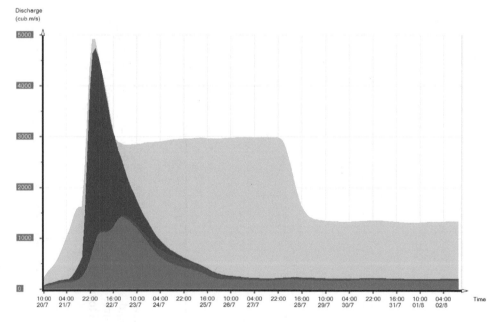

Fig. 6.14. Huayuankou hydrograph with Xiaolangdi's release added up on top (green coloured area)

138

6.5.3 Exchange of ideas with relevant users

During the research, preliminary results and conclusion have been presented during the "YRCC Internal Seminar" (in English) in March 2010 and August 2011. The internal seminars were held for showing the latest research results by YRCC staffs and for training the people's skills to join English-based international conferences. Experts and colleagues agreed with the main conclusions and were very interested in the implication for future development. Not only the colleagues on flood control department showed the interests in the research result, but also those from the water resources management and allocation department appreciated the output. The later suggested that the finding should be used to guide the future development of water resources management system, which are also based on the same multi-reservoir-based regulation river system, but operated in non-flood season.

6.6 Computational technologies for next generation software systems

Since computational speed is an important factor in the decision making process, different ways to achieve this are explored below.

6.6.1 Supercomputing

Supercomputing has been involved in the highly calculation-intensive tasks such as problems including quantum physics, weather forecasting, climate research, oil and gas exploration, molecular modeling, and physical simulations. The real-time reactive operation demands very fast computing capacity. It requires much more computations than the normal input-process-output execution. Supercomputing makes possible to run big 2D simulation models at much faster speed. The YRCC has set up its supercomputing center in 2006 in an effort to upgrade the computing capacity as required by large simulations and highly calculation-intensive researches.

At present, the center serves mainly large simulation models in hydraulic research. Suggestions have already been made to run operational systems with need of big computing capacity, including the 2D modeling of flood in lower Yellow River.

6.6.2 Pre-computed scenarios and historical reference base

For a big basin like the Yellow River, various spatial and temporal distributions of rainfalls lead to enormous flood patterns in terms of locations, sizes and hydrographs. At the same time, each flood has its own developing process. Different operational alternatives may lead to many model simulation runs which might be time consuming. Therefore, the method of pre-computed scenarios can be a feasible solution. This has been studied, by researchers, for the decision support purpose. Chailan et al (2012) introduced the technique of pre-computing scenarios instead of simulating online to provide a service in the common objective of understanding quickly and accurately what the live conditions are and what will be their consequences. This is to get rid of the dependencies on these constraints to provide a forecast once a crisis arises (Chailan et al, 2012). The conclusion is that the concept of pre-computed scenarios can be used for alert system tools.

6.6.3 Training ANN on pre-computed results

Artificial Neural Network (ANN) is a powerful computational tool having the capability of capturing underlying characteristics of any physical process from the dataset and is able to extract the relation between the input and output without the underlying physics being explicitly explained. The application of ANN on calculation-intensive should be explored for flood control and management decision support in providing the pre-computed simulation results.

6.7 The 3Di Water Management project

A recent development in The Netherlands for developing a better tool is the '3Di Water Management' project. Delft University of Technology, Deltares and Nelen & Schuurmans are the initiators and developers of the 3Di Water Management System. Delfland Water Board, Water Board Hollands Noorderkwartier and Water Platform Haaglanden are the launching customers of the project. Water Board Vallei and Eem and Water Board Veluwe are involved as supporting customer. 3Di Water Management is a four-year innovation program in which various (IT) products are being developed for water managers, specially developers and emergency organizations. These products emphasize a faster and more accurate prediction of possible flooding. According to the developers, 3Di Water Management offers: (i) detailed information about flooding as a result of heavy rainfall and floods; (ii) immediate insight into the effects of measures; (iii) real-time information, accessed via an interactive web portal; (iv) clear insight in flooding by means of realistic 3D-animations (3Di website, 2013). Here, the 'immediate insight' and 'Real-time information, accessed via an interactive web portal' are the effort towards the good management tool.

A big challenge is to use very detailed digital elevation maps (with a resolution of 0.5 x 0.5 m) for detailed flood computations on a large scale. While existing mathematical flood models can deal with schematization of 1 million cells, the 3Di algorithm currently deals with schematizations consisting of more than 1000 million cells. Normally models of this scale would require calculation times of days, weeks or months. With the newly developed flexible mesh technique, however, computation times can be restricted to minutes or hours (van Leeuwen and Schuurmans, 2012).

In addition to the development of advanced 3Di algorithm, the Technical University of Delft developed a 3D environment based on the mapping of detailed elevation data, in which the results of high-resolution flood simulations can be projected.

In future there will be a 3D stereo visualization of floods with their motion processes. Another interesting innovation is the possibility of interactive modeling. Because the presentation of simulation results and model adaptations can be made during the actual simulation, the tool is especially suitable for decision support in calamity situations and design table situations. With three mouse clicks, the dikes can be raised while the effect on the course of the flood is visualized immediately. In projects on land-use planning, the possible interaction and advanced visualization can be used to show policymakers, governments and water boards the effects and consequences of improvement works.

Especially for the identification of climate adaptation measures such as green roofs and other forms of retention measures, interactive modeling has proved to be very effective. And it is also a way of enthusing the public. The tool will contribute to the public being progressively more aware of water management problems and possible solutions (van Leeuwen and Schuurmans, 2012). These elements show exactly the main focus towards the new software applications, as has been called the next generation software in our research. The new effort is extremely important for a successful management and decision support system, which is goal of this research.

6.8 Conclusions

From functional composition point of view, the multi-reservoir-based flow regulation process can be designed as a three-layer structure, which is rather similar to the software environment that has been proposed by Mynett and de Vriend (2005). The multi-reservoir-based flow regulation is designed to have three layers. They are the bottom-level components, reservoir-layer components and basin-layer functions.

Engineering-based flood control and management can greatly benefit from DSS software applications that support the decision making process. However, despite these potential benefits there have been many reports that DSSs are hardly used in practice, since they do not appeal to end-users' ways of operation.

The system architecture as well as the execution pattern of DSS systems should be designed in accordance with and based on well understanding of the actual decision making process.

Introducing the concept of a real-time reactive mechanism has shown to be a good choice for developing such systems. In such approach, the model/simulation component is embedded in an event-driven operational loop rather than following a procedural programming based input-run-output approach. This enables a highly efficient and intuitive approach that is very much in line with the actual decision support process.

Decision makers as well as collective decision meetings are quite special users of software applications. They always require high-level presentation of information. An innovative approach for designing a flood control and management DSS user interface is to follow the technical tradition of working with profession-based display graphs for the system. Examples in flood control include hydrographs, rating curves, water levels, storage volumes, and release time series for reservoir operation.

The two application examples presented here show the potential of introducing the real-time reactive mechanism in DSS design. The single-reservoir application was perceived by the test group of the Flood Control Office of the Yellow River Conservancy Commission as having a very good future. The example for multi-reservoir operation decision shows even more practical support for decision making process which is based on more complex computation in which reservoir regulation, flow routing in river channels and time series summing are made to form a complete reservoir-river system computation.

What have been given in this chapter are examples of application on some key stages of decision making process. More functions relating to decision support should be developed for a more complete flood control and management decision support system.

Advanced DSS and application of new technologies: nowadays, it is possible to have advanced DSS in terms of fast speed and real-time reactive operation. The supporting technologies may include supercomputers, check scenario's against experience database, pre-computed scenarios in model database, emulation control points by training ANN on pre-computed results.

The "3Di Water Management" project in the Netherlands emphasizes a faster and more accurate prediction of flooding possible. Its achievements will be offering immediate insight into the effects of measures; real-time information, accessed via an interactive web portal; clear insight in flooding by means of realistic 3D-animations (3Di website, 2013). This provides an advanced solution for obtaining quick response of concerned information which, in our new approach, plays an important role.

7 Case study applications

In this chapter the proposed DSS approach is applied to two case studies on flood events: one for a past flood event on the Yellow River and a second one for a projected flood under climate change conditions. The adaptive management and decision making process is shown in the presented case studies along with all the necessary explanations of the applied rules and the analysis of the operational decision. For the historical case, the rules are introduced and the general process of developing alternatives is presented. The second case study application uses a large flood based on historical records but with a higher flood peak and higher discharge values. This is to show the possible effect of climate change on flood events - a similar discharge volume but more concentrated in time. The case application is made to emulate real-time operations with special focus on how the alternatives are formulated and optimized.

The case study applications also help to show the general process on how the proposed alternatives are evaluated and optimized in real-time and in a man-machine interactive way. The resulting operating rules for the reservoirs are given and explained for the chosen case studies. The flow calculation schematization for the river-reservoir system is as presented in Chapter 6. The calculations were based on the prototype software tools developed during this research.

7.1 Yellow River flood control rules and operational conventions

The general target of the reservoirs regulation is to operate the four reservoirs collaboratively to regulate and optimize the flow downstream in order to eliminate the possible loss or reduce them to a minimum. Based on this principle, the rules for reservoir operation, established by YRCC are:

(1) Rules for all four reservoirs:

- The flood season normal level (FSNL) should be kept as the maximum operational level during non-flood time and should be the target level during flood season when flood regulation operation applies.

- In general, the operational starting level is supposed to be equal to the flood season normal level; with special situation, such as the case with an extreme flood forecasted, the starting level can be lower than the FSNL. In this case, a pre-release operation may have been made for the propose to extend the flood regulation pool.

- In case a flood is routed through a reservoir, as the overall flood process is approaching the end, the reservoirs should discharge the storage volumes that are above their flood season normal levels as soon as possible. This demands the reservoirs to be prepared for possible future floods. Meanwhile, if no urgent or extreme condition such as the dam in danger, the releases should not form big flow which may lead to loss in the downstream of the reservoir. Releases should be made to reduce the Huayuankou hygrograph's peak discharge and its high-flow part as small as possible, if the release is able to have influence on the peak.

- Operation of a reservoir should be done in conjunction with the other reservoirs in such a way that a reasonable hydrograph is obtained at Huayuankou, so that losses are reduced to their least. When the rainfall is forecasted to decrease firmly and the flood is going to an end, the operation should be to discharge all reservoirs storage above the flood season normal level to the after-peak part (tail part) of the flood at Huayuankou station in a safe way for the entire river system. Namely, this operation should not form big flow which may lead to more loss in the downstream of the reservoir.

- During flood regulation period, flood control and management has the top priority. Generation of power and other beneficial operations should be made under the condition that will not lead to more flood threat.

- With all things considered, if an urgent situation occurs, special action can be taken based on decision made from formal procedure and approved by relevant authority. (An example of the situation occurred in the downstream of Guxian reservoir was that a few workers were stranded by flood water in a bridge construction field. Guxian's release was required to reduce so as to give the condition for rescue workers to get close to the pier where the construction workers could only stay on the flat top. Then the reservoir had to reduce its release for some hours to enable the rescue operation. Urgent situations like this example are actually quite popular in real-time operations).

(2) Rules for the Luhun and Guxian reservoirs:

- Releases from the reservoir should be no bigger than 1000 m3/s, (the safe discharge for both reservoirs' downstream channels), if the operational level is below the flood regulation limitation level (FRLL) and above the flood season normal level (FSNL).

- Maximum release should be made, if the operational level is equal or bigger than the flood regulation limitation level. In this case, the safety of the dam becomes a main concern.

(3) Rules for Sanmenxia reservoir:

- Except for the extreme flood from the upper stream, the reservoir is not allowed to retain flood water. The flood season normal level will be kept for general floods. (The function to regulate an incoming flood has been radically reduced from its original design. This is due to the reason that, after the reservoir was put into operation, it was noticed that higher storage level causes big sedimentation in the reservoir's backwater area and a higher river bed may lead to severe upstream overland flow or even dike failure to the nearby region. Thus, operational principles have been greatly changed from the design).

(4) To carry out practical alternatives, the following operational conventions are generally used:

- **Forecast definition in Yellow River multi-reservoir river system:** The Yellow River flood control strategy is to define forecast as the flow in case that the reservoirs keep the "present" operation status or there is no flow regulation by the reservoirs. More specifically, the forecast result is based on the condition that, for each reservoir, the outflow hydrograph equals to the inflow hydrograph. So, the forecast focuses on rainfall-runoff and river flow predictions.

- **Advance time:** this is the approximate time length for a discharge value or a section of a reservoir's release to affect a target value or section of the resultant hydrograph in downstream. Fig. 7.1 gives a simple example showing the time for a flood to go from the upper stream to the downstream hydrographs of a flood: (i) red line, the upper boundary hydrograph; (ii) blue line, lower boundary hydrograph. "t1" is the time needed for peak to travel from upper to lower boundary. In YRCC, this advance time is a practical and experiential parameter for operators to plan and implement the release for a downstream control target hydrographs. It is the time needed for a peak or a point at the outlet hydrograph to travel to the point of the control target hydrograph. A more accurate method is to link the advance time with different magnitude of discharge values or even the hydrograph shapes. For example, if the peak discharge is around 4000 m3/s, then 8 time steps (16 hours) is used as the advance time for the peak section of hydrograph, whereas 6 time steps (16 hours) for 6000 m3/s. If a unique value is used, then it is usually an average and statistical value from previous floods. In our case, the advance time for Xiaolangdi's release is set to 7 time steps, namely 14 hours. Compared to number of subsections of the Muskingum parameter, 6, it is approximately one time step more or 2 hours more. A longer advance time will make more accurate control of the downstream hydrograph. However, the advance time also has a close link with the lead time of flood forecast. In general, the former should be equal to or less than the later.

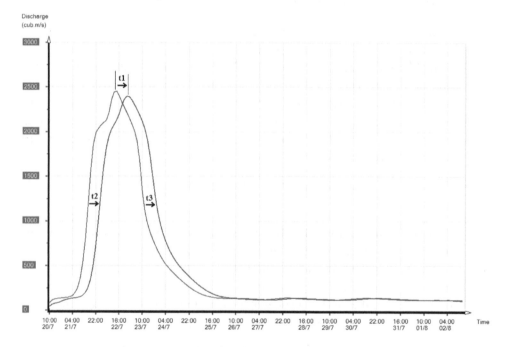

Fig. 7.1. Hydrographs of upper and lower boundaries

- **The control target discharge at the Huayuankou station:** As discussed in Chapter 4, the present smallest bankful discharge is thought to be around 4000 m3/s. If release from reservoirs will make the Huayuankou's hydrograph 'fat', then the discharge, if possible, should be equal or less than 4000 m3/s. The peak section of the hydrograph can last rather long time, but the discharge value should be around 4000 m3/s rather than bigger. A smaller control target discharge will also be safe for the downstream floodplains, but the time for the reservoir to return to FSNL will be longer. And this can only be allowed under particular conditions. For example, forecast shows no upcoming flood for enough long time, thus the preparation of the flood reservation pool is not so urgent.

- **The release hydrograph:** if the hydrograph from Xiaolangdi's release is used to control or optimize the downstream target hydrograph, then it has to be calculated or estimated before the actual release starts. The release hydrograph and the downstream control target hydrograph have a complicated relationship for two major reasons: first, the release hydrograph will be routed to the downstream (in either hydrological or hydrodynamic way). In the routing process the hydrograph attenuates and this varies from one hydrograph to another, with different shape and magnitude of the hydrograph. The new (routed) hydrograph will be part of the downstream resultant hydrograph. Therefore, the following assumption is used to calculate/estimate the required release hydrograph: the attenuation is small enough so that it can be neglected. Or in another word, the hydrograph needed by the control target hydrograph at downstream is used as the release hydrograph after the advance time is considered. This means that the needed hydrograph is made to be the required release hydrograph simply by "shifting" the time in the scale of advance time.

- **Lead time:** in real-time operation, the lead time of forecast has to be considered. At YRCC, 8 hours is the usual lead time for the hydrologic bureau to issue a "standard forecast" for Huayuankou station. "Standard forecast" refers to forecast which meets national accuracy standard, including peak discharge error < 20%. Usually the standard forecast is a "short-term forecast". In addition to the standard forecast, another forecast may be used for reservoir operation, the mid-term early warning forecast which is less accurate but has a longer lead time up to 20 hours. According to the YRCC internal protocol on flood forecasting, on the mid-lower Yellow River, there are three different forecast results for Huayuankou station. They are "standard forecast", "reference forecast" and "alert (forewarning) forecast". They have different lengths of lead time and different service purposes.

Usually, standard forecast is used for operation of reservoirs. Reference forecast is used for preparation of flood control and relevant management actions. Alert (forewarning) forecast is used for informing relevant stakeholders and governmental departments so as to get things prepared for possible flood. If necessary, reference forecast is also used for decisions of reservoir operation. For river flow forecast by routing, the error should be less than 20% for peak discharge increment. For example, the prediction is the peak discharge will increase by 2000 m3/s, then the error should be less than 400 m3/s (for rainfall runoff forecast, the error should be less than 20%).

7.2 Case study 1 - historical flood

In this case study, a historical flood is taken as an example to show the application of the proposed solution. The flood is a small one and therefore floods in this magnitude happen more frequently. In the past, operational management and decision making was supported by generally a simulation system or a so-called decision support system but with the same execution pattern of a simulation system. An operational scenario or alternative will be first suggested and then translated into input data for the simulation system. The simulation system will be run by the people who support the flood control discussion or decision meetings. Due to the inadequate execution time and style, bad user interfaces, this run was not allowed be done in the flood control room or the decision room. Thus, the simulation output would be delivered to the discussion meeting or decision meeting. This output was usually simulation result based on the proposed alternatives. Sometimes, a few variations of alternatives might also be simulated to provide comparison. Then they would be presented in the meeting for the decision makers and experts to analyze. If some changes were made to the alternatives, then more runs would be needed again. This process took lot of time. The decision makers were not satisfied with it and also the simulation model operators were busy and tired to response to request and run the simulation. The support for the what-if based analysis was not in a direct fashion and thus was inefficient.

In this case study application, the proposed real-time reactive operation based on new system architecture and man-machine interaction is used to support the management meeting or decision meeting. The run of the tool will directly serve the meeting for the development or evolution of the alternatives. Suggestions and ideas will be responded with no delay in real-time. Detailed description is shown in the following case study application.

7.2.1 General information of the 1996 flood

The flood here is based on a flood event in July and August 1996. It is a relatively small flood with origination in both upper stream of Sanmenxia reservoir and the Sanmenxia-Huayuankou region. The return period of the flood is smaller than 5 years in terms of the peak discharge of the natural runoff recorded at the Huayuankou hydrologic station (YRCC, 2009). These kinds of floods are usually called, by YRCC, normal floods, and they have gained more and more attention due to the ever growing requirements and standards for protection. At the same time, improved engineering, like the operation of

Xiaolangdi reservoir, provides conditions for controlling not only big floods but also the more frequent small floods as well. Therefore, operational scenarios are also made for mid and small floods in an effort to upgrade floods with different magnitudes.

As has been discussed in chapter 6, the given "forecast" is 11 discharge time series: 10 local inflow hydrographs of the 10 sub-catchments in Sanmenxia-Huayankou area and 1 inflow hydrograph from upper stream of the Yellow River main stream (see the 11 ellipses in Fig. 6.4). Table 7.1 gives the general information about them.

Time information of the given time series is:

- Starting time 10:00 o'clock 20 July 1996

- Ending time 16:00 o'clock 02 August 1996

- Time step length: 2 hours

Table 7.1 The small flood in case study 1

No.	ID	Time series sources - contributing regions and Yellow River trunk	Peak discharge (m^3/s)	Total volume ($10^6 m^3$)
1	Q01_Gu_in	Inflow of Guxian reservoir	192	39.13
2	Q02_Gu_Chang	Guxian-Changshui local inflow	228	44.43
3	Q03_Chang_Yi	Changshui-Yiyang local inflow	256	47.83
4	Q04_Lu_in	Inflow of Luhun reservoir	1762	185.37
5	Q05_Lu_Long	Luhun-Longmenzhen local inflow	534	65.41
6	Q06_Yi_Bai	Yiyang-Baimasi local inflow	324	61.46
7	Q07_B.L._Hei	Bai.-Long.-Heishiguan local inflow	494	83.36
8	Q08_Xiao_Hua	Xiaolangdi-Huayuankou local inflow	3860	352.92
9	Q09_San_Xiao	Sanmenxia-Xiaolangdi local inflow	384	81.16
10	Q10_San_in	Inflow of Guxian reservoir (from upper stream of Yellow River trunk)	3173	1672.32
11	Q11_(Qin)_Wu	Flow from Qinhe river branch at Wuzhi	146	45.31

From Table 7.1, it can be seen that the runoff coming from the upper main stream Q10 of the Yellow River has a bigger volume and quite a big peak discharge. The flow from upper main stream and Xiaolangdi to Huayuankou local inflow (rainfall runoff) Q08 will be the two dominative sources for the resultant hydrograph at Huayuankou hydrologic station. Based on the rules, the predicted Huayuankou hydrograph is the main factor for determining an operational flood control action in the mid-lower Yellow River. The peak discharge of Huayuankou is composed from combined flows of the upper stream Yellow River and the entire area of Sanmenxia to Huayuankou region. Fig. 7.2 and Fig. 7.3 show the inflow hydrographs into Luhun, Guxian, Sanmenxia and Xiaolangdi reservoirs.

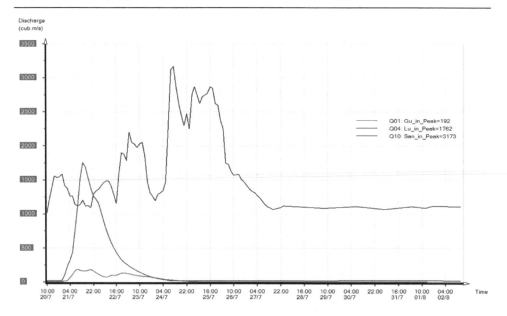

Fig. 7.2. Inflow hydrographs into the Guxian, Luhun and Sanmenxia reservoirs

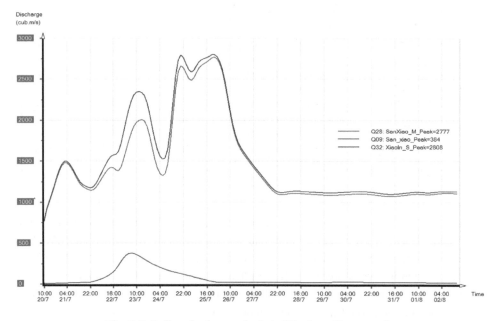

Fig. 7.3. Inflow hydrographs into Xiaolangdi reservoir

Sanmenxia reservoir inflow hydrograph, shown in Fig. 7.2 in purple line, is the given forecast of the main stream of the Yellow River, upper stream of Sanmenxia. Both Luhun and Guxian reservoirs have closed catchments, therefore they have simple inflow hydrographs obtained from rainfall runoff models, shown in blue and red lines in Fig. 7.3. Xiaolangdi reservoir's inflow consists of flow routed from Sanmenxia outlet and the rainfall-runoff forecast of the local region, or sub-catchment, between Sanmenxia and Xiaolangdi (see lines in Fig. 7.3 where the red line is the inflow from Yellow River main stream and, blue line, the local inflow by rainfall runoff; purple line, sum of the two as total inflow).

7.2.2 Reservoir operation in the proposed DSS

Based on the general operation rules and the operational target, several some basic alternatives can be developed as a starting point for making better and operational alternatives. Studying the basic alternatives helps in finding the main issues that should be focused on dealing with the floods. Modification to the basic alternatives will generate new ones which will be more practical. This is the general evolution process of an alternative towards a better one.

Therefore, in order to get a general picture about a flood, "default" alternatives are proposed, in which special assumptions are made, for example, no regulation of flood (i.e. outflow is equal to inflow), utmost regulation of flood (all inflow stored in reservoirs and release equals to zero). The next step is to modify these alternatives to be reasonable and operational for the given flood. Fig. 7.4 shows five alternatives where the first two are based on special assumptions, which are definitely not the feasible solutions for operation but useful references for people to get the general information about the flood.

Here is a description of the considered alternatives:

(1) Natural flows - no regulation

In this alternative, each reservoir's outlet will discharge equal to the inflow in the reservoir. In general, this makes the reservoirs to function at minimum releases. Alternatively it can be said that the effect of this alternative is the same as if all the dams were removed. The flow in the river is a "natural flow". It can help us understand the magnitude of the flood before any human intervention.

The resultant hydrograph at Huayuankou hydrological station, for this alternative is shown in Fig. 7.6 (Q34). The peak discharge is 6636 m^3/s, which is a value higher than the critical value, 6500 m^3/s, of the bankfull discharge, in the section near Huayuankou. Since further downstream the river has a smaller bankfull discharge (the smallest is around 4000 m^3/s,), the flood is obviously a threat to some of the lower floodplains. Therefore, reservoir regulation is necessary.

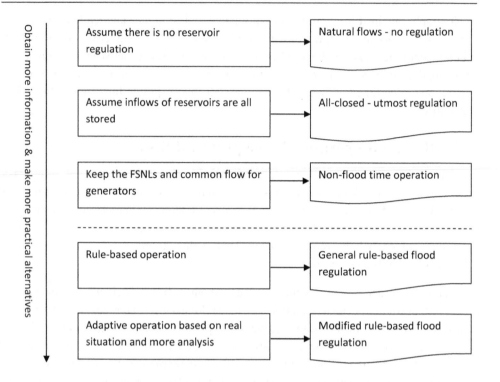

Fig. 7.4. Alternatives for analysis and decision making

SOBEK 1D2D model simulation was run in order to represent the flood extent in the downstream. Experience of other researchers shows that it is important to first look at the maximum flooding extent, to determine the flood risk (Popescu et al., 2010; Van et al., 2012).

The result obtained, when the Huayuankou hydrograph was considered as the upstream boundary input is shown in figure 7.5, for the moment of the biggest flooding extent. In this map, it can be seen that a subsection of the simulation domain has overland flooding at the edges of floodplains on both sides of the main stream.

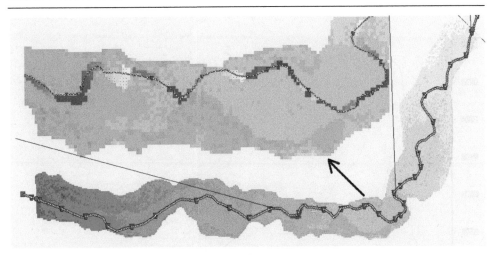

Fig. 7.5. SOBEK 2D overland flow simulation result with the natural-flow alternative

(2) All reservoirs closed - fully regulated

In this alternative, all outlets (including generators) are closed, which means that all values of a release are made equal to zero. This is an assumption which makes the reservoirs to capture all the inflows. In this alternative it is assumed that the reservoirs are able to store all the inflow volumes. Therefore at Huayuankou observation point, all flood water is due to the rainfall runoff of the area downstream of the, Xiaolangdi, Luhun and Guxian reservoirs. The flood inflow to Xiaolangdi, including that from Sanmenxia's release and that from Sanmenixa to Xiaolangdi local inflow, will be stored by Xiaolangdi.

The resultant hydrograph at Huayuankou hydrologic station is shown in blue line in Fig. 7.6 where the hydrograph (red line) of no-regulation alternative is also shown for a comparison. The peak discharge of the blue line is 4643 m^3/s that is much smaller than 6636 m^3/s. Since these two alternatives correspond to two extreme operational schemes, these two peck discharges represent the extreme values that can be obtained, i.e. the biggest peak discharge 6636 m^3/s and smallest 4643 m^3/s. More practical alternatives of optimal functioning of reservoirs would lead to peak discharges between the two numbers.

SOBEK 1D2D model simulation result obtained in this second alternative is shown in Fig. 7.7. There is no flooding of the floodplains. This is reasonable, for the smallest bankfull discharge in the SOBEK simulation domain is 5000 m^3/s, a value bigger than the peak discharge of the hydrograph. However, the all-closed alternative requires that the reservoirs capture all their inflows and therefore it might put the reservoirs at risk in case of big flood events. In most cases, it is not a practical solution for general flood operation.

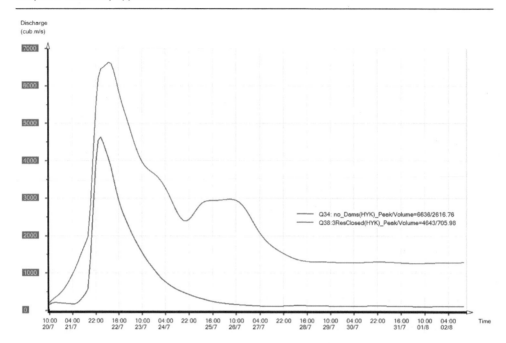

Fig. 7.6. The Huayuankou hydrographs

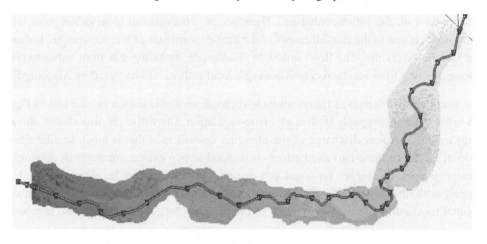

Fig. 7.7. SOBEK 2D overland flow simulation result with the all-closed alternative

(3) Keep the FSNLs and common flow for generators

This third alternative is the general operation in flood season with normal flow or small flood events. According to the general rules presented in section 7.1, the initial storage levels are all set to be the FSNL. The initial values for river flow routing are all set to be the first values of upper boundary inflow time series.

Table 7.2 shows the storage levels. It can be seen that Luhun is the only reservoir whose storage level goes higher than the FSNL, but the highest level is 318.21m which is much less than the designed flood regulation limitation level (FRLL, 323.00 m).

Table 7.2 Design and operational feature levels

Operational levels (m)	reservoirs			
	Xiaolangdi	Guxian	Luhun	Sanmenxia
Initial level =FSNL	225.00	529.30	317.30	305.00
Highest level	225.00	529.30	**318.21**	305.00
Ending level	225.00	529.30	317.50	305.00
Designed maximum flood regulation level	275.00	548.00	323.00	335.00[*]

*Note: the Sanmenxia reservoir can only be put into flood regulation operation under extreme flood conditions due to its severe sedimentation problems. In general, the FSNL will be its operational level during flood regulation process.

Fig. 7.8 shows the difference, in which: (i) red line is the release hydrograph which is equal to inflow; (ii) blue line is the release hydrograph 'no bigger than 1000 m^3/s'.

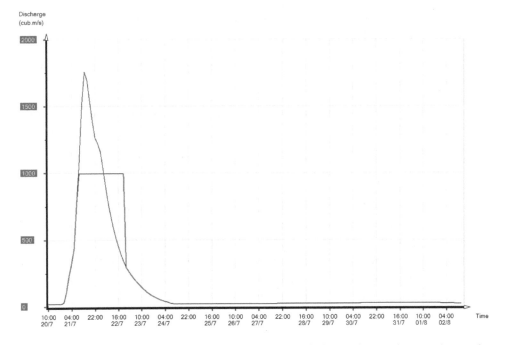

Fig. 7.8. Luhun reservoir's releases in case of no regulation and a maximum release of 1000 m^3/s

The SOBEK1D2D simulation is almost the same as in Fig. 7.7 (and therefore it is not shown here). The reason for this similarity is that we have the same flood volumes and slight different peaks and shapes. The difference is too small for a substantial change of the 2D simulation results.

(4) Rule-based alternative

As required by the rules, one important operational goal is that, with the regulation of the reservoirs, the big-flow part of the hydrograph at Huayuankou is made as small as possible and the flooding threat to the floodplains is reduced to minimum. In the same time, if a reservoir's storage level gets higher than the FSNL, it should realease in such a way that it will return to the FSNL as quick as possible, in order to get the flood reservation pool fully resumed, to be prepared for future floods. In our case, that the biggest peak discharge at Huayuankou station is 6636 m^3/s (without any reservoir regulation) is representative for a small flood event. In general, such magnitude of floods leads to no dam safety concern. Therefore, the main operation should be focused on reducing the peak section discharge and optimized the hydrograph so that the downstream flooding threat can be reduced as much as possible.

In order to find the required hydrograph to control the downstream hydrograph "no bigger than 4000 m^3/s", the following simple calculation is made:

For each (time step) value of downstream summation hydrograph (summation excluding hydrograph from Xiaolangdi's release),

- if it is bigger than 4000 m^3/s, let release equal to zero; otherwise, let the value equal to the difference between the value and 4000 m^3/s;

- the time for each value is shifted by the advance time (e.g. 7 time steps).

After these two steps, the required release is obtained with correct time indications. In Fig. 7.9, the hydrograph in red line (hydrograph Q29) is the required release hydrograph obtained with the above-mentioned processing. The one in blue line (hydrograph Q30) is the resultant downstream hydrograph that is obtained by routing from the upstream dam site to downstream point. The deformation can be seen clearly in the figure. The peak discharge is lowered while the zero-value discharge section of the release has a corresponding section, in the resulting hydrograph, which is bigger than zero.

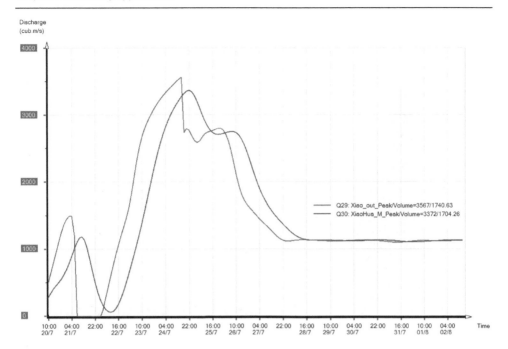

Fig. 7.9. Required hydrograph (Q29) for Xiaolangdi's release operation and the downstream corresponding routing result (Q30).

The resulting hydrograph at Huayuankou station is shown in Fig. 7.10, where the after-peak flow is controlled to be around 4000 m³/s when Xiaolangdi reservoir discharges its stored flood volume to get the flood reservation pool ready for future floods. The numbered time series Q26, Q31, Q08 and Q30 refer to information of normal hydrographs while the peak values marked with different colors are overlapped hydrographs - the outer line of the green area is the Huayuankou's hydrograph. It can be seen that the peak and its nearby section are reduced remarkably as a result of Xiaolangdi's release. The peak discharge is 5379 m³/s. This is 237 m³/s bigger than the summation of the hydrographs Q26, Q31, Q08 which are three flow sources besides Xiaolangdi's release. The overlapping display of these three hydrographs is shown in Fig. 7.11 where Q26 is from Yiluo river, Q31 from Qinhe river and Q08, forecasted runoff of Xiaolandi to Huayuankou subcatchement. After the release scheme is decided for Guxian and Luhun, the hydrographs shown in Fig. 7.11 will be fixed. It can be regarded as a sort of known conditions. Thus, the Xiaolangdi's release becomes a variable to determine the final shape of the Huayuankou hydrograph.

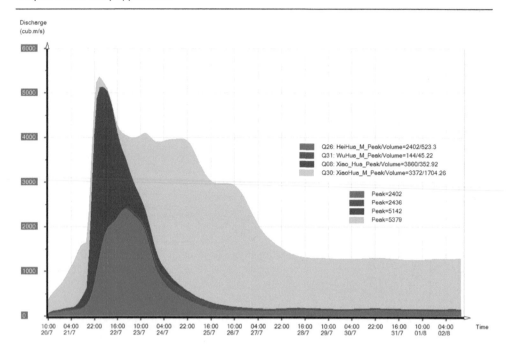

Fig. 7.10. Formation of the hydrograph at Huayuankou station

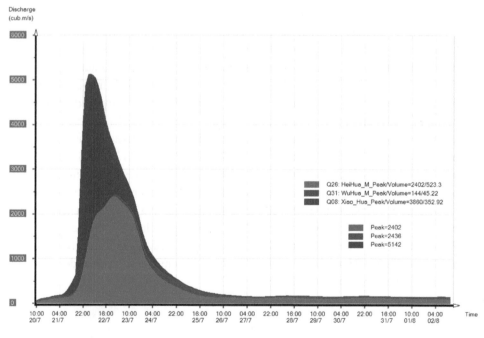

Fig. 7.11. The overlapping display of hydrographs for flow Q26, Q31, and Q08.

7.2.3 Selection of the operational alternative

The target hydrograph of a regulation operation, the resultant hydrograph of Huayuankou, can be thought as consisting of two parts, one which is changeable and one unchangeable. The changeable part is the part which is formed from the release of reservoirs. This is the result of the reservoir regulation and reservoirs' downstream river routing. The unchangeable part is a combination hydrograph of all runoff from the sub-catchments below the reservoirs, namely, they cannot be regulated by the reservoirs. The optimum of the resultant hydrograph is done by changing the regulation schemes of the reservoirs. Floods originated from upper stream of reservoirs would allow more capability for regulation than those from downstream. At the same time, flow from reservoirs need time to reach the operational target points. This time will also be a factor affecting the control of the target hydrograph. For example, if a peak needs to be lowered at a time 8 hours later, then the reservoir for the control should be Xiaolangdi, rather than Luhun and Guxian. For the later two reservoirs' releases "travel" more than 20 hours to get to Huayuankou and therefore are not able to control the target time point.

An alternative can be modified or optimized towards the users' preference with all possible controlling factors stated above. Adaptive management and decision making would be supported with the efficient control of these factors and real-time reactive presentation of control result. In our case, flood mainly comes from the upper stream of Xiaolangdi reservoir. By adjusting the release of Xiaolangdi, different Huayuankou hydrographs can be produced. For example, the release for the after-peak part of the hydrograph is set to be 3000 m^3/s (shown in Fig. 7.12). If the two resultant hydrographs of Huayuankou are compared, it can be seen that with 4000 m^3/s Xiaolangdi can reach its FSNL in a shorter time. But the downstream will get bigger flow. If there is forecast that a following flood (in the near future) is developing, then a quicker release might be the decision makers' choice. A satisfactory alternative can be obtained with the help of efficient decision support by changing the release hydrograph of Xiaoangldi and observe the behavior of Huayuankou's hydrograph.

In this application case, the inflow flood is relatively small. So the highest storage level reached during the regulation is not the main concern of the operation. If the flood is quite big, a high storage level may require the change of operation based on the rules. Also, in real-time operation, any emergent situation could be an important issue for making a decision.

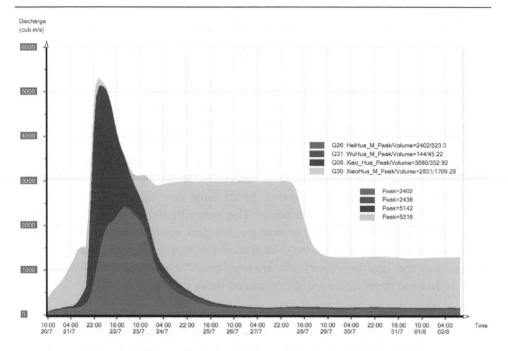

Fig. 7.12. The overlapping display of hydrographs for controlling Huayuankou hygrograph's peak section to its least and making the after-peak part equal to 3000 m³/s

The management issues presented in this case study are elaborated together with the explanation of how the decision making process can be efficiently supported using the proposed tool. In the past and also at present with the software tools still in use, the process is quite different. These software tools do not allow real-time operation during the decision meeting, but are performed somewhere else, for example in the technical people's office or computing room. This is a very slow process. When compared with the proposed real-time reactive operation as presented here, a big step forward can be made. As mentioned in Chapter 6, the YRCC is very interested in such new tool. Suggestions have been made to build a complete operational system based on the proposed approach.

7.3 Case study 2 - flood under climate change conditions

7.3.1 General information on the flood

The flood in this case study is based on a historical flood with peak discharge 11400 m³/s, but this flood is made more concentrated in time, maintaining the same total volume but shortening the duration which leads to a higher peak discharge (see Fig. 7.15 for comparison of the two hydrographs). This change could create a flood which is more

160

threatening to the downstream safety, representing possible future effects of climate change. In this way, extreme conditions can be emulated and corresponding management options can be studied in a virtual computer-based environment. The hypothetical flooding event in this case study corresponds to a return period of about 50 years, based on design flood analysis (YRCC, 2010).

The original historical event caused big loss of properties of people living in the floodplains. At that time, there were only two reservoirs in the Sanmenxia-Huayuankou region and Xiaolangdi and Guxian dams had not been built yet. Sanmenxia is on the upper stream of the flood originating area, so it was not able to regulate the runoff yielded in this area. Luhun got intensive flood in short time and therefore played important role in regulating the flood at Huayuankou. At the same time, there was also heavy rainfall below the reservoirs, so the uncontrolled rainfall-runoff also accounted for a major part of the Huayuankou hydrograph. This limits the regulative effect of reservoirs to the entire flood. This section presents the general process for flood control decision support based on our solution.

The given forecast consists of 11 time series: 10 local inflow hydrographs of the 10 sub-catchments in the Sanmenxia-Huayankou area and 1 inflow hydrograph from the Yellow River main stream (see Fig. 6.4). Table 7.3 gives the detailed information about them. The main information of the given time series is:

- Starting time 10:00 o'clock 20 July 2005
- Ending time 16:00 o'clock 02 August 2005
- Time step length: 2 hours

Table 7.3 The various contributions considered in case study 2.

No.	ID	Time series sources - contributing regions and Yellow River trunk	Peak discharge (m^3/s)	Total volume $(10^6 m^3)$
1	Q01_Gu_in	Inflow of Guxian reservoir	2181	233.4
2	Q02_Gu_Chang	Guxian-Changshui local inflow	1047	112.6
3	Q03_Chang_Yi	Changshui-Yiyang local inflow	3654	356.7
4	Q04_Lu_in	Inflow of Luhun reservoir	6778	642.6
5	Q05_Lu_Long	Luhun-Longmenzhen local inflow	5130	380.2
6	Q06_Yi_Bai	Yiyang-Baimasi local inflow	1877	270.6
7	Q07_B.L._Hei	Bai.-Long.-Heishiguan local inflow	1154	152.1
8	Q08_Xiao_Hua	Xiaolangdi-Huayuankou local inflow	2997	761.3
9	Q09_San_Xiao	Sanmenxia-Xiaolangdi local inflow	4671	907.9
10	Q10_San_in	Inflow of Guxian reservoir (from upper stream of Yellow River main stream)	pm	pm
11	Q11_(Qin)_Wu	Flow from Qinhe river branch at Wuzhi	1990	819.6

From Table 7.3, it can be seen that the flood is mainly from the Sanmenxia-Huayuankou area. The inflows into Luhun (Q04) and Xiaolangdi (Q08) are quite big in terms of flood volumes. Meanwhile, relatively big flood volumes are also seen in the areas downstream of reservoirs, including Sanmenxia-Xiaolangdi (Q09) and Xiaolangdi-Huayuankou (Q08). Fig. 7.13 and Fig. 7. 14 show them in graphs.

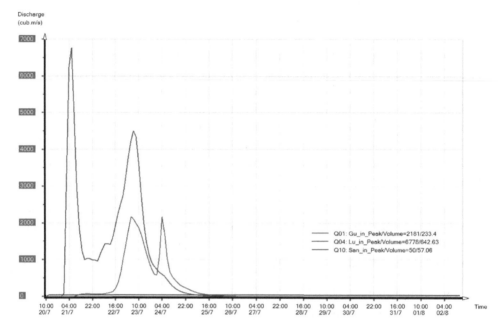

Fig. 7.13. Inflow hydrographs of reservoirs Guxian, Luhun and Sanmenxia

In Fig. 7.13, the red line is the inflow of Guxian reservoir, and, the blue line, Luhun inflow, purple line, Sanmenxia inflow. Fig. 7.14 shows Xiaolangdi's inflow hydrographs: the red line is the inflow from Sanmenxia's release, and the blue line, local inflow from the Sanmenxia-Xiaolangdi region, the purple line, sum of the two as a whole of Xiaolangdi's inflow.

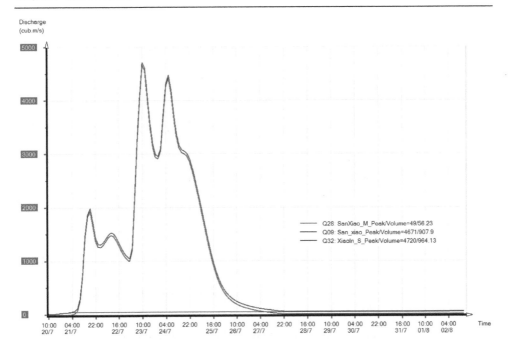

Fig. 7.14. Inflow hydrographs of reservoir Xiaolangdi

7.3.2 Preliminary analysis of the flood

With a big flood, the general target of the regulation is to operate the four reservoirs collaboratively to regulate and optimize the flow to downstream in order to eliminate the possible loss or reduce it to its minimum. Before making any decisions, some basic alternatives will again be to provide general ideas of the flood.

(1) Natural flows - no regulation

Initial values: (i) Yellow River main stream is kept outside the calculation for the tributaries; (ii) other river sections: the first values of upper boundary inflows.

The resultant hydrograph at Huayuankou hydrologic station is shown in Fig. 7.15 (Q34). The peak discharge is 19478 m^3/s which is much bigger than the critical bankfull value, 6500 m^3/s, of the overland flow in the section near Huayuankou.

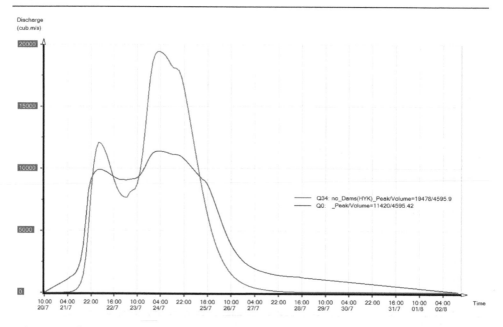

Fig. 7.15. The natural flow hydrographs (assume there is no reservoirs): Q34, more time-concentrated with a bigger peak; Q0, the original flood

The Huayuankou hydrograph was used as an upstream condition for the SOBEK 1D2D. Fig. 7.16 shows the moment of the biggest flooding. From this map, it can be seen that overland flooding occurs widely along the river. Only few pieces of lands are not inundated for their high elevation. If this happened in reality, there must be big loss of properties in the floodplains.

Fig. 7.16. SOBEK 2D overland flow simulation result with the natural-flow alternative

(2) All gates closed - fully regulated

Initial values: (i) Yellow River main stream: 50 m³/s; (ii) other river sections: the first values of upper boundary inflows; (iii) reservoirs' initial storage levels (take their FSNLs): Luhun, 317.50 m; Guxian, 529.30 m; Sanmenxia, 305.00 m; Xiaolangdi, 225.00 m.

The resultant hydrograph at Huayuankou hydrologic station is shown in blue line in Fig. 7.17, where the hydrograph (in red line) of no-regulation alternative is also shown for a comparison. The peak discharge of the red line is 19478 m³/s which is much bigger than 12484 m³/s. Since these two alternatives correspond to two extreme operational schemes, the two peak discharges represent the extreme values that can be obtained, i.e. the biggest peak discharge 19478 m³/s and smallest 12484 m³/s. More practical alternatives based on rather optimal functioning of the reservoirs would lead to peak discharges between these two numbers.

The SOBEK 1D2D model simulation is shown in Fig. 7.18. Compared with Fig. 7.16, the flooding area in Fig. 7.18 is much smaller. However, with this flood, inundation is unavoidable due to the fact that this flood is quite big and that runoff generated downstream of reservoirs accounted for certain part of the entire flood. The alternative of all outlet closed has given rather clear idea about unavoidable flooding area.

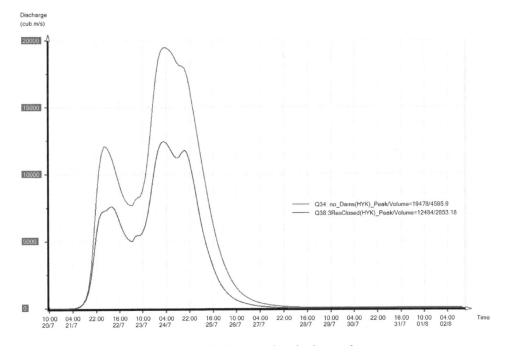

Fig. 7.17. The Huayuankou hydrographs

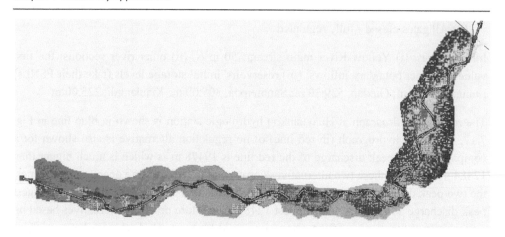

Fig. 7.18. SOBEK 2D overland flow simulation result with the all-closed alternative

7.3.3 Rule-based alternatives and decision making

The magnitude of the flood has been shown with the above two alternatives. Then more alternatives can be obtained towards practical solutions. The rule-based alternatives can be made based on the rules and considerations of real-time situations. Variations can be made to the operations to make different alternatives.

For big floods, reservoirs should be used to reduce the Huayuankou's peak discharge section as much as possible with the consideration of hydropower generation. Whether the generators are allowed to be operated will be decided by decision makers under the real-time conditions of the floods. According to the rule, Xiaolangdi's power generation is usually given a discharge of about 1000 m^3/s for three generators with each approximately 300 m^3/s.

According to the Yellow River flood control rule, if the forecast suggests that the peak discharge of Huayuankou is bigger than 10,000 m^3/s, then the Xiaolangdi's release can be made to control the Huayuankou's flow rate equal to or less than 10,000 m^3/s during the after-peak time when it discharges its storage volume above the FSNL. At the same time, if the prediction is that there is no flood in near future, the reservoirs might be allowed to release in a smaller discharge. This would give the reservoirs longer time of operation in above the FSNL. Based on different considerations, various alternatives can be produced for the search of satisfactory operational choice.

Again, during a decision meeting, quick response of the software tool is very important for supporting the decisions. The decision makers can change the release hydrograph very easy with graphical operation while at the same time, the resultant Huayuakou's hydrograph will response in real-time reactive mode. Whenever a change is given, a updated resultant hydrograph will be presented immediately without delay. For the decision making process, this would be much more efficient over the traditional IPO or normal simulation based systems.

In Fig. 19, the purple colored hydrograph (Q37) is the Huayuakou hydrograph being controlled by Xiolangdi's release with a target control discharge 10,000 m³/s. This hydrograph is already a reasonable and rule-based alternative. If no urgent discharging is needed, a smaller control discharge can also be considered.

Fig. 7.19 shows the four hydrographs: (i) Q34, for the case of natural flow (assuming there are no reservoirs); (ii) Q38, for the case of all reservoir sluices closed (no release); (iii) Q37, both Luhun and Guxian's releases are equal to or smaller than 1000 m³/s, Xiaolangdi's release is in two sections: peak section, no release; after-peak section, control the Huayuankou's flow rate equal to or less than 10,000 m³/s; (iv) Q35, for the case to keep FSNL.

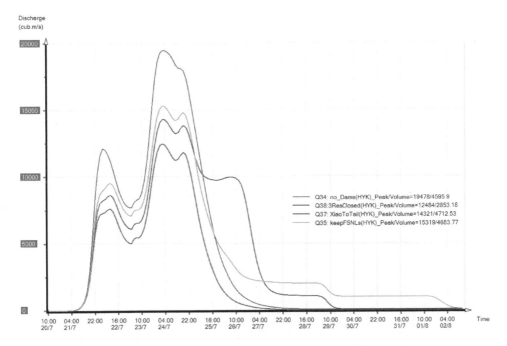

Fig. 7.19. Comparison of Huayuankou's hydrographs of different alternatives

Fig. 7.20 shows the resultant hydrograph with it's various contributions. The upper edge can be seen as the purple line Q37 in Fig. 7. 19. In Fig. 7.20, the purple area represents the flow from the Yiluo branch of Yellow River where Luhun and Guxian locate. There the purple area is the regulation result of both reservoirs. It can be seen that this flow contributes significantly to the peak section of Huayuankou hydrograph. In general, both reservoirs can be made to release less in order to make the peak section smaller. But both reservoirs releases need more than 20 hours to reach Huayuankou which is much longer than the formal forecast which is usually officially taken as 8 hours under the condition that there is rainfall runoff below the reservoirs. Therefore, in real-time operation, the two reservoirs cannot be used to precisely control the Huayuankou hydrograph. That is why their releases can be allowed to be 1000 m³/s during the peak section of the resultant hydrograph in our alternatives. The 1000 m³/s limit is based on the two reservoirs downstream river conditions in terms of safety. For all reservoirs, the highest storage levels reached during the regulation are below the flood regulation limitation level (FRLL), therefore, dam safety is not a concern in making a decision.

With support of adaptive management tool, the decision makers can explore alternatives to optimize decisions. Decisions will be made based on real-time conditions. The 1D2D overland flow simulation is suggested to be made fast enough to be taken into the decision loop by taking advantage of new technologies which have been listed and discussed in Chapter 6.

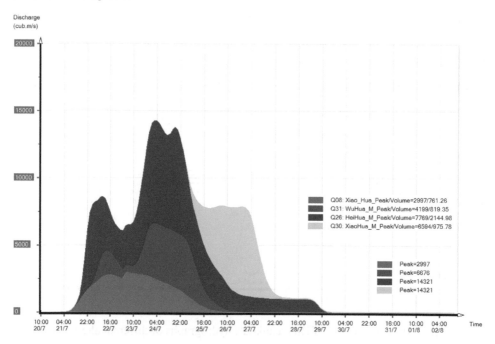

Fig. 7.20. The resulting hydrograph for controlling Huayuankou hygrograph alternative

In Fig. 7.20 the resulting hydrograph follows from adding the varioous contributions: both Luhun and Guxian's releases are equal to or smaller than 1000 m³/s, Xiaolangdi's release is in two sections: peak section, no release; after-peak section, controlled by Huayuankou's flow rate equal to or less than 8000 m³/s

7.4 Discussions

7.4.1 The process of obtaining a good solution

From the two examples of this chapter, the following general process was used to provide a practical and direct working way towards more acceptable alternatives:

- Work out few reference alternatives as a base. This is done to provide the decision makers a general idea about the incoming flood, including its magnitude, its distribution in space and time, the extent to what the reservoirs can regulate the flood, etc.

- Making more realistic alternatives based on ideas obtained from the general alternatives. In this case the decision makers can have ideas for further improvement. Make the necessary changes to the general alternatives with the efficient support of the software tool, more practical and operational alternatives can be made and presented.

- Making final decisions. By discussions and comparisons, a final operational alternative will be determined. In this process, efficient support tool is very helpful for the decision making process.

7.4.2 Reservoirs' releases and downstream resultant hydrograph

In our research area, there are four reservoirs. The first considered reservoir is Sanmenixa which is working in series with the downstream Xiaolangdi reservoir. Sanmenxia's release will flow into Xiaolangdi and will not have a direct influence on the hydrograph of Huayuankou.

The other three reservoirs are Guxian, Lulun and Xiaolangdi, which are responsible for the control of the target hydrograph, at Huayuankou station. However, due to their different locations and sizes, these three reservoirs will have different power to control the hydrograph. Besides the flood regulation capacity, their locations are a major factor determining their contribution to the downstream hydrograph, for a location will decide the time needed for a release flowing to the downstream control target point (see Fig. 7.21).

Fig. 7.21 shows simple sketch of the idea. At the time simultaneous, the releases from the three reservoirs will contribute to different parts of the resultant hydrograph due to the different distances to the target point and different river flow conditions. Of the three reservoirs, Xiaolangdi is much nearer to Huayuankou, so its release will affect the resultant hydrograph in much shorter time.

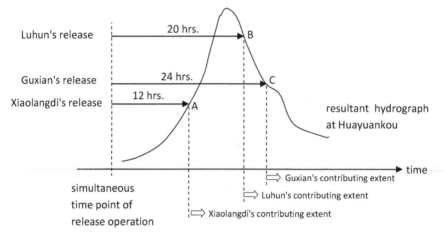

Fig. 7.21. Reservoir release times contributing to the downstream resultant hydrograph

Based on statistics of historical events and flow simulations, the approximate time needed for the releases to travel from the dam sites to Huayuankou are obtained as basic information for reservoir operation in accordance with lead time of real-time forecasting. Release from Xiaolangdi needs the shortest time, approximately 12 hours, to reach Huayuankou. Therefore, Xiaolangdi reservoir has both bigger regulative capacity and favorite location   for controlling the resultant hydrograph.

7.4.3 Support for decision making

The relationship of the outlet hydrograph and the control target hydrograph is quite complex due the fact that the three reservoirs' outlet hydrographs need to be routed to the downstream and in the same time they join with other lateral local runoff and then form the resultant target hydrograph.

The isolation of the required hydrograph makes it possible for the decision making process to focus on the key issues, such as the control of the peak discharge and peak-section discharge, the control of inundated areas in the floodplains by means of 1D2D simulations. Then, the use of real-time reactive operation will provide very efficient decision support (Li and Mynett, 2011). The target hydrograph of operation, the resultant hydrograph of Huayuankou, can be thought as consisting of changeable part and unchangeable part. The changeable part is the part which is from the release of reservoirs. The changeable part is the result of the reservoir regulation and reservoirs' downstream river routing. The unchangeable part is combination hydrograph of all runoff from sub-catchment below the reservoirs, namely, they cannot be regulated by the reservoirs. The optimization of the resultant hydrograph is done by changing the regulation schemes of the reservoirs. So, floods originated from upper stream of reservoirs would allow more capability of regulation than those from downstream. Adaptive management should provide a direct and efficient way to produce a hydrograph which meets the decision makers' idea. The manipulation of release hydrographs will lead to the resultant hydrograph to change in real-time reactive manner. This makes the decision support process extremely efficient, flexible and visualized.

Therefore, a successful decision support tool means not only the necessary functional deployments of the system, but also how the system serves its user, how the system interacts with its user.

In the two applications, only some general aspects are mentioned to compare different alternatives. This research only shows some of the key features that an adaptive tool should own. In a real-time operation, more factors may be considered to make a decision. Urgent engineering situations, socio-economic factors, even political issues can be a concern for making decisions. Therefore, the decision making is an issue related to many real-time conditions. Then, the more complex the issue is, the more necessary the adaptive management tool is needed.

Inflows of the two examples are all within the regulation ability of the reservoirs. If they are too big, then the reservoirs may reach very high storage level. In this case, the dam safety would become essential for the reservoir operation.

7.4.4 Other issues

Due to the heavy sediment load, sediment is a main issue for the Yellow River. Present research however does not address this issues because (i) in times of floods, the management of water quantity is usually urgent and a priority; (ii) managing flow and sediment together requires a proper sediment modeling or flow-sediment transport coupling modeling tool. And reliable modeling tools are still needed for sedimentation modeling for both reservoirs and river channels. At present, sediment simulations for evaluating an alternative cannot be made into the decision making process, due to its low accuracy for sediment transport, sedimentation and morphological changes, long execution time, complexity of initial and boundary conditions and difficulty of coupling with flow simulation.

The YRCC is currently making effort towards a quasi real-time sediment simulation manner for the evaluation of flow managing alternatives in the real-time operation decisions (Zhu, 2005; YRCC, 2006; Li, 2009). In this manner, only few alternatives, e.g. those closely related to sediment, are required to run the sediment simulations.

7.5 Summary

In this chapter, two applications are presented to show the effect of developing alternatives for reservoir operation in order to reduce the downstream flooding extent. In combination with the proposed adaptive real-time reactive management and operation approach outlined in Chapter 6, a very direct and efficient way can be developed to explore alternatives more efficiently and support the decision making process.

Interactively adapting the release hydrographs of the various reservoirs involved will change the resultant hydrograph in a real-time reactive manner, making the decision support process more efficient and flexible. The use of 1D2D inundation simulation models allows evaluating alternative decisions in a more direct and visualized way. More factors may have to be considered in real-time operation and decision making. E.g. any urgent situation on dame safety can be a main concern for making decisions. A complete system based on the proposed approach is suggested for real-time operational management and decision at the Yellow River Conservancy Commission.

8 Conclusions and Recommendations

8.1 The Yellow River and floods

The Yellow River is sometimes called 'the cradle of Chinese civilization' since it runs from the Himalaya's to the Bohai Sea and this is where ancient Chinese civilization developed. Spiritually, the river nurtures the entire Chinese nation. However, the Yellow River is also said to be the 'China's Sorrow' or the 'Scourge of the Sons of Han' for historically its major floods often have had catastrophic effects on people and land. The 1887 flood devastated the area and was said to have caused more than 900,000 dead. In 1933, around 18,000 people were killed by less devastating, but still widespread flooding, while in 1938 a flood was triggered by Chinese Nationalists deliberately in an attempt to disrupt the Japanese invasion, with hundreds of thousands of people losing their lives.

Since ancient times, flood control and protection has always been an important task for the people living along the Yellow River. The heavy sediment load, which largely originates from the easily erodable Loess Plateau in the middle section of the river, is responsible for the fine silt loading in the lower Yellow River. In history, sedimentation caused continuous rising of the downstream river bed and thus led occasionally to complete changes in the river planform and often to catastrophic flooding. The Yellow River Conservancy Commission (YRCC), a special branch of the Ministry of Water Resources of China, was established in 1946 to manage the Yellow River, one of its major roles being flood control and management. The YRCC acts as the Yellow River Flood Control and Drought Relief Commanding Headquarters.

8.2 The operational process and new requirements

During the past sixty years, at least ten large multi-purpose dam projects were constructed by the YRCC in the main stream of the river, including the Sanmenxia and Xiaolangdi reservoirs. Many other large reservoirs were built in the tributaries of the Yellow River as well, such as the Guxian reservoir on the Luohe river, and the Luhun reservoir on the Yihe river, and on many other branches in the middle reach of the river. Among all of them, the four reservoirs Sanmenxia, Xiaolangdi, Guxian and Luhun play a dominant role in the flood control and management of the mid-lower Yellow River.

The entire river system with its tributaries and multiple reservoirs, constitute a complex network for storing, regulating and conveying flows. In such system, the operation of reservoirs is essential for the control and management of floods and the protection of people and property. The YRCC has gained considerable experience with the operation of this multi-reservoir river system. Collective decision making meetings with high ranking officials and experts are responsible for developing operational scenarios, which are of utmost importance as a flood develops under the ever changing reservoir storage conditions and varying river flood conveying status.

Any decision making process involves many exchanges of information and computations, including analysing meteorological information and weather predictions, flood forecasting reports, status of reservoirs, operational objectives, and assessing proposed measures, possible hazards and possible alternatives. All in real-time operation.

In YRCC as well as in other large river authorities in China, the analysis and decision making is usually done in collective meetings. These meeting are joined by decision makers, technical experts, staff in charge of flood forecasting, hydrologic information, engineering management, etc. In these meetings, all the latest issues that are key to the flooding situation will be studied and examined before decisions are made.

- The general process in flood management meetings of Chinese river authorities is that the latest information on hydrological forecasts, reservoir conditions and other relevant information are presented and discussed in a high-level meeting. After the overall goals are set, more specific problems are focused on, such as which reservoirs should be operated and how to regulate the flood. Existing decision support tools are used to categorize the effects of possible operational alternatives. The less important issues will be dealt with by technical experts, while the more crucial decisions are discussed and agreed collectively in the decision making meeting.

The what-if discussion and analysis need to be supported efficiently in real-time, for no delay is acceptable in a crucial decision meeting with limited forecast lead time. The computational time of any software tool is crucial for providing efficient support to the decision making process. Once decision makers are clear about the overall goals, they need near-real-time interactive verification of operational alternatives.

Real-time flood control and management also implies that a decision meeting could be held at any moment as the flood develops. A potential flooding event could involve a number of meetings during different flood development stages.

A solution worked out automatically by a computer can only be a reference for the decision makers meetings. Decisions have to be based on analysis of real-time situations. From an adaptive management point of view, no computer can easily work out an ideal solution for them. Therefore, the mechanism to quickly interact with a DSS system for developing alternatives is the most important function of any flood control and management software system.

Since the 1980s, research and development has been carried out at YRCC to support optimal reservoir operation. However, after many years, many decision makers and end-users of such systems are still not satisfied. The systems are often not considered efficient and user-friendly in supporting the decision making process. Investigations at other large river authorities in China gave similar conclusions.

- Obviously, there is a need to develop new software tools for more efficient support of the decision making process which invariably takes place under time constraint.

8.3 Multi-reservoir operations and requirements

From the practice on the Yellow River, it is found that 'controlled release' is a very common regulation alternative for real-time flood control and management. Chow (1964) stated that if the gates are operated during the routing period, (i.e. controlled release operation), then the routing process easily becomes complex. Controlled release has not drawn enough attention though it is a frequently used regulation process. Multi-reservoir-based flow regulation on the Yellow River requires special reservoir routing computations. In general, reservoir routing is based on mass conservation of in-pool storage water. From this point, it should be a simple process. However, the Yellow River operational rules involve various computations based on different release schemes which cannot be dealt with directly by traditional methods, which are usually based on uncontrolled releases and single-valued functional relationships between release and storage. The multi-reservoir regulation on Yellow River demands not only uncontrolled but also controlled release schemes, including constant release for minimum water level variation.

- Different parts of a multi-reservoir river system may have quite different storage-release relationships which require correct and robust routing techniques.

In the Yellow River flood control operation, the typical controlled release regulations may contain one or more of the following: (i) release in accordance with a pre-defined hydrograph; (ii) maintaining a given storage level; (iii) applying different release policies for different reservoir levels; (iv) imposing a constant release discharge for a given storage variation; (v) allowing a variable release discharge based on inflow or downstream conditions. All these controlled release regulations require more a complex computational scheme in comparison with a simple uncontrolled release procedure.

- It is important to delicately deal with the transition from one particular release scheme to another, because mass conservation has to be strictly maintained; software tests at YRCC show that computational algorithms need to be improved.

Complex operational management rules or scenarios lead to complex routing calculations. A typical example is maintaining a particular target reservoir level during operation. This regulation calculation is a complex rule-based controlled-release routing process. When the level is below the given target level, it is generally required to return as quickly as possible for power generation purpose. When the level is above the given level, it should be made to return to the level while the flood regulation function is implemented. So, if the level is below the given level, the release is set to be either zero or at a minimum value. If the level is higher than the given level, the reservoir could be in operation of a flood regulation process. This implies that

- a robust algorithm and program implementation is needed.

Maintaining a given target level regulation is a rule-based water mass conservation calculation. The traditional method cannot fit different rules around the target level (maintaining at the same time maximum level for flood prevention versus minimum level for hydropower can lead to oscillations around the target level, as sometimes observed).

8.4 Advances in numerical flood simulation

Simulation models are basic tools to evaluate the impact of different reservoir simulation alternatives. They should also be a basic component of the management and decision support tool. Different ways to operate the reservoir system of the Yellow River may have different inundation effects on the lower Yellow River floodplains.

- Nowadays advanced numerical simulation models are available. In this thesis, the prediction of the inundation pattern on the Yellow River was done using the state of the art simulation tools produced by Deltares Software Center, in particular SOBEK 1D2D and the new D-Flow Flexible Mesh *βeta* version.

These two tools use different solution approaches, viz. a uniform grid and a flexible grid resp., in order to build the simulation models which can be used for flood management and decision support. The two models show the general principle of how alternatives can be evaluated by assessing the inundation effects of different reservoir operations. More work is needed to incorporate these models in real-time operational tools with much smaller cell sizes and adequate calibration, but

- the feasibility of using present-day advanced numerical simulation tools rather than a target hydrograph is clearly shown in this thesis.

Both uniform grids and flexible grids were used to discretize the simulation domain. With uniform grids, the domain consists of cells all having exactly the same size of square or rectangular shape. Flexible grids allow variable cell shapes within the domain. This helps focus more on the representation of the flow fields based on gradients in the terrain, which allows a reduction of the total number of cells. Due to limited accuracy of DEM and lack of relevant data for the Yellow River, no particular emphasis was put in this thesis to the accuracy of the model. But the principle has clearly been shown.

8.5 A flexible method for reservoir routing

Reservoir routing is the basic process of calculating the flow from runoff through a reservoir. A robust algorithm and a flexible software application are needed when dealing with different rules that depend on the target water level. Traditional routing methods usually work under the condition that the release is a simple function of water level or storage (uncontrolled release). However, a release scheme may be a combination of particular prescribed release conditions (controlled release). There are basically two types of controlled release operations: (i) the release and storage level have a one-to-one relation, i.e. every water level (or storage volume) will correspond to only one release discharge value; (ii) the release and storage level can have multiple relations, which means that the release is not only decided by the water level (or storage volume) alone, but also by other factors, such as upstream inflow or downstream outflow.

- In order to achieve flexible reservoir operating capabilities, in this thesis a new method was developed to deal with 'multi-valued controlled storage-release relationships' that assures the routing process to be both accurate and robust. The application condition is generic in that the storage-release relationship only needs to have some mathematical 'relation' and is not limited to a single-valued function.

The new method, called Cross Line Method (CLM), is based on the known characteristics of the reservoir and the (multi-valued) relationship between storage and water level, or storage and release. Compared with other methods, CLM has more flexibility and potentially less error. The method obtains the storage volume and release discharge by locating the intersection of the V-q relation for mass conservation with the V-q relation for the reservoir's characteristics.

177

In the Cartesian plane, the intersection coordinates are the V-q values for that particular time step. The storage level can then be obtained by interpolating the reservoir's characteristics using V. So, this method seeks the solution based on the discrete data each time step. It is simple and accurate, and the algorithm can be easily run on a desktop computer. It is suitable for a wide range of applications as long as the reservoir release has some known relation with storage or storage level. Hence it can easily be implemented in a multi-reservoir regulation software tool.

8.6 Software architecture for flexible decision support

Different available software tools for reservoir operation were studied in this research. It was found that different systems or applications require different architectures. As an example, the Delft-FEWS system developed by Deltares is a software tool based on a *data-centered* architecture, rather than on a traditional *model-centered* approach (Werner et al, 2004). Various utilities are made available to deal with generic processing of data in the context of flood forecasting that interface with the database, as well as an open interface to the modeling systems that effectively allows incorporation of a wide range of forecasting models, independent of the supplier. A *data-centered* approach allows flexible use of open shell architectures.

- Traditionally a *model-centered* IPO (Input-Process-Output) architecture is often used for model-based simulation systems. However, nowadays a *data-centered* approach facilitates a more flexible architecture which allows model components to be easily updated or interchanged. From a functional decomposition point of view, multi-layered architectures can be used to serve end-user applications.

The software architecture suggested by Mynett and de Vriend (2005) can be applied to all relevant fields of river management with the multi-reservoir-based flood management application being one of its components. Such system would be especially suited for flood management and decision. It allows a functional decomposition which also focuses on decision support.

After examining a number of systems in the fields of flood control and flood management decision support, it was found that researchers and software developers often build a decision support system in much the same way and same style as numerical modeling and simulation systems. In this case, the simulation model is at the center of the process while the system architecture follows the 'batch-like' execution procedure of the model. This can be referred to as a 'simulation-centered' architecture. These types of systems are very popular for model-based applications in research, such as hydrological or hydrodynamic modelling systems. However, investigations on the requirements formulated by end-users

178

of Decision Support Systems show that an approach is favoured that supports optimizing and improving alternatives in a fast and interactive way. Such approach was explored in this thesis (Li and Mynett, 2011):

- the concept of a 'real-time reactive mechanism' has shown to be a good choice for developing decision support systems. In such approach, the simulation model component is embedded in a user-driven operational loop rather than following a program-driven input procedure. This enables a highly efficient and intuitive approach that is very much in line with the actual decision support process.

The reactive system piloted in this thesis is purely input-driven and can react at a pace dictated by the end-user. A real-time reactive system continuously reacts to stimuli received from the environment by sending messages, where each reaction must be made within certain time bound (Berry, 1999; Wan, 2002). Real-time, reactive programming seems a most appealing basis to create a really advanced and yet practical DSS system for collective flood control discussion and decision making meetings, in view of its generic relation with the underlying decision making process. As basic support, a functional decomposition was developed that enables flexible computation of optimal multi-reservoir regulation. This three-layer structure is similar to the proposed software environment by Mynett and de Vriend (2005) for more general applications at YRCC, covering not only flood control but also optimal water management.

The user interface design and interactive operation modes are also major factors for a successful DSS. In accordance with present day software systems, professional graphical user interfaces are proposed to provide intuitive man-machine interaction in a real-time reactive manner. It is suggested that low-level access to data must be avoided and that information should be presented in a graphical way that is easily understandable.

Pilot studies carried out in this research with various user interface designs and interactive operation modes for both single and multi-reservoirs, show the feasibility of this approach to support an actual decision making process. The proposed solution demands fast response of the software tool, which nowadays has become possible, as shown by the 3Di research project in The Netherlands (3Di Waterbeheer, 2013). The supporting technologies may include using supercomputers, checking possible scenario's against an experience database, using pre-computed scenario's in a model database, or emulating particular control points by training ANNs on pre-computed results.

- A next generation DSS should allow for these options.

8.7 Case studies for the mid-lower Yellow River basin

Decision making for flood control in the Yellow River basin is aimed at optimizing operational alternatives. Applying the real-time reactive approach in a case study for the mid-lower Yellow River proved very efficient to quickly identify the target hydrograph for operating the Huayuankou reservoir. The hydrograph can be separated into a changeable part and an unchangeable part. Optimization is done by changing the reservoir regulation schemes and observing the simulated flooding. The use of 1D2D inundation simulations and advanced visualization techniques allows evaluating alternatives in a more direct and physically understandable way.

- An adaptive real-time reactive management approach can provide direct and efficient ways to explore alternatives and efficiently support the decision making process.

The case study applications show some of the key features that an adaptive tool should have. In real decision making more factors may have to be considered like structural integrity of the dams, socio-economic factors, political issues, etc. A complete DSS system for real-time operational management should include these features as well.

8.8 Recommendations

The framework developed in this thesis is intended to show the generic principles of adaptive management, but of course should be further developed before it can be applied in actual operations. Experts from YRCC that were consulted about the pilot versions, have suggested that a more complete system with all necessary user interfaces should be developed based on the finding of this research to serve the Yellow River flood control and management. The numerical modeling of the lower Yellow River should be given much more attention to obtain really practical solutions for real-time operation. The simulation time should be as short as possible so as to achieve near-real-time interactive decision support, like the 3Di pilot system in The Netherlands.

The proposed real-time reactive approach can be applied to other YRCC software tools as well. One immediate suggestion is to apply this concept in hydrologic modeling. If the calibration of the underlying system can be done in a real-time-reactive manner, it could radically improve the time and effort involved, while reservoir operators can target their operations in a very efficient and easily understandable way.

REFERENCES

Allan, C., 2007. *Adaptive management of natural resources,* Proceedings of the 5th Australian Stream, Wilson, A.L., Dehaan, R.L., Watts, R.J., Page, K.J., Bowmer, K.H., & Curtis, (Eds), Management Conference. Australian rivers: making a difference. Charles Sturt University, Thurgoona, New South Wales.

Ashworth, M. J., 1982. *Feedback Design of Systems with Significant Uncertainty,* Research Studies Press, Chichester, UK.

Balica, S. F., Popescu, I., Beevers, L., Wright, N., G., 2013. *Parametric and physically based modelling techniques for flood risk and vulnerability assessment: a comparison,* Journal of Environmental Modelling & Software, 41(3), p. 84-92

Betancourt, J. L., 2009. *Adaptive Management of Climate Change Impacts on Ecosystems: Some Personal Perspectives.* Adaptation Conference archive at web: http://www.law.arizona.edu/depts/ele/adaptiveconferernce (accessed Oct. 2012)

Berry, P.M., Drabble, B., SWIM, 1999. *An AI-based System for Workflow Enabled Reactive Control,* in proceedings of the IJCAI Workshop on Workflow and Process Management held as part of IJCAI-99, Stockholm, Sweden, August 1999.

Brodie, R., Sundaram, B., Tottenham, R., Hostetler S., Ransley T., 2007. *An Adaptive Management Framework for Connected Groundwater-Surface Water Resources in Australia,* Bureau of Rural Sciences, National Landscape Program, Australia

Burton, I., 1997. *Vulnerability and adaptation response in the context of climate and climate change. Climate Change,* by Kluwer Academic Publishers. 36, 185-196

Chailan R., Bouchettey F., Dumontierx C., Hessx O., Laurent A., 2012. *High Performance Pre-Computing: Prototype Application to a Coastal Flooding Decision Tool,* (author manuscript) published in 2012, Proceedings of the 4th International Conference on Knowledge and Systems Engineering, Da Nang, Viet Nam"

Changjiang (Yangtze River) Water Resource Commission (CWRC), 2006. *Development Plan for Changjiang Flood Control System* (No. YTFCS-DSS-DD-01) (in Chinese), National Flood Control and Drought Relief Commanding System (Project Phase I), 10-11

Chow, V. T., 1964. *Handbook of Applied Hydrology,* A Compendium of Water-resources Technology, McGraw-Hill Book Company, 1964

Cluckie, I., 2012. *The Flood Risk Management Research Consortium,* FRMRC website, http://www.floodrisk.org.uk/ (accessed Nov. 2012)

Cui, J., Xin, G., Cai, L., 1998. *Research and Development of Yellow River Flood and Ice Control Decision Support System,* First Edition, Yellow River Press, ISBN: 9787806211441. 78-79

Custer, S. G., Sojda, R. S., 2006. *Adaptive management of water resources AWRA summer specialty conference*, Missoula, Montana, AWRA 26-28 June

Dahm, R., 2006. *Usefulness of flood forecasting decision support systems: A brief report*, UNESCO-IHE course: Flood modeling for management

Delta Program website, 2012. *Delta Programme*, http://www.deltacommissaris.nl/ (accessed Dec. 2012)

Deltares, 2012. *RIBASIM, River Basin Planning and Management*, Deltares Innovative solutions for water and subsurface issues. http://www.deltares.nl/en/ software/479948/ribasim2 (accessed Nov. 2012)

Deltares website, 2012. *What is 3Di Water Management?* http://www.deltares.nl/en/ product/ 1767530/what-is-3di-water-management (accessed Dec. 2012)

Dessai. S., Hulme, M., Lempert, R. J., Pielke, R. J. 2009. Climate prediction: a limit to adaptation? In: Adger WN, Lorenzoni I, O'Brien K (Eds) Adapting to climate change: thresholds, values, governance. Cambridge University Press, 64-75

Dessai, S., Hulme M., 2007. *Assessing the robustness of Adaptation decisions to climate change uncertainties: A case study on water resources management in East of England,* Global Environmental Change, 17,59-72

Dinh, N.Q., Balica, S., Popescu, I., Jonoski, A., 2012. *Climate change impact on flood hazard, vulnerability and risk of the Long Xuyen Quadrangle in the Mekong Delta,* Journal of River Basin Management, 10 (1), 103-120

DHI, 2006. *The Lemhi River Mike Basin Model: A Tool for Evaluating Stream Flows, Diversion Operations and Surface Water-Ground Relationships in the Lemhi River Basin,* Idaho, DHI, Inc.

Dhondia, J. F., Stelling G. S., 2002. *Application of one dimensional - two dimensional integrated hydraulic model for flood simulation and damage assessment,* Proceedings of the Fifth International Conference on Hydroinformatics, Cardiff, UK

Engle, N. L., Johns O. R., Lemos M. C., Nelson D. R., 2011. *Integrated and Adaptive Management of Water Resources: Tensions, Legacies, and the Next Best Thing.* Ecology and Society 16(1), 19

Fenton J. D., 1992. *Reservoir Routing,* Hydrological Science Journal, 37 (3), 233-246

Flood Control 2015 website, 2012. http://www.floodcontrol2015 .com /about-flood-control-2015 (accessed Nov. 2012)

Frank, E., Ostan, A., Coccato, M., Stelling G. S., 2001. *Use of One Dimensional and Two Dimensional Hydraulic Approach for Flood Hazard and Risk Mapping,* Proceedings of the 1st Conference on River Basin Management, R. A, Falconer and W. R. Blain (Eds.) , WIT Press, Southampton, UK, 99-108

Gao, K., Liu, Y., Dou, S., Yang, G., 1991. *History of Yellow River Flood Control,* Vol. III (in Chinese), Henan Public Press, ISBN: 7-215-01434-7, 19-56

Gheorghe, S., 2008. *Adaptive management of the climate change problem: bridging the gap between research and public policy*, Munich Personal RePEc Archive, MPRA. Http://mpra.ub.uni-muenchen.de/13564 (accessed Oct. 2012)

Gichamo Z., G., Popescu, I., Jonoski, A., Solomatine D.P., 2012. *River Cross Section Extraction from ASTER Global DEM for Flood Modeling*, Journal of Environmental Modelling & Software, 31 (5), 37-46

Gray, P., 2005. *Adaptive Management in the 21st Century, Parks Research Forum of Ontario (PRFO) State-of-the-Art Workshop*, Monitoring in Ontario's Parks and Protected Areas

Hassaballah, K., Jonoksi, A., Popescu, I., Solomatine, D. P., 2012. *Model-based optimization of downstream impact during filling of a new reservoir: case study of Mandaya/Roseires reservoirs on the Blue Nile River*, Water Resources Management, 26(2), 273-293

Hillborn, R., 1992. *Institutional learning and spawning channels for sockeye salmon (Onchorynchus nerka)*. Canadian Journal of Fisheries and Aquatic Science, 4, 1126-1135

Holling, C.S., 1978. *Adaptive Environmental Assessment and Management*, Published by Wiley, Chichester, UK. ISBN 0-471-99632-7.

ICFM5 website, 2012. *5th International Conference on Flood Management* (ICFM5), http://www.ifi-home.info/icfm-icharm/icfm5.html (accessed Feb. 2012)

Information Center of Ministry of Water Resources (ICMWR), China, 2006. *Construction Plan for the National Flood Control System,* The Project Book of National Flood Control and Draught Relief Commanding System (in Chinese), Ministry of Water Resources, China, 3-4

Intergovernmental Panel on Climate Change (IPCC), 2007. *Climate Change 2007: Synthesis Report*, IPCC, 30-50

International Flood Initiative (IFI) website, 2011. *International Flood Initiative*, http://www.ifi-home.info/ (accessed May 2012)

Jonoski A., Popescu I., 2011. *Distance learning in support of water resources management: an online course on decision support systems in river basin management,* Water Resources Management, 26(5), 1287-1305

Kahan J. P., Wu M., Hajiamiri S., Knopman D., 2006. *From Flood Control to Integrated Water Resource Management - Lessons for the Gulf Coast from Flooding in Other Places in the Last Sixty Years*, http://www.rand.org/content/dam/rand/pubs/ occasional_papers/2006/RAND_OP164.pdf (accessed Jan. 2013)

Keen, P.G.W., 1997. *Let's focus on action not information*, Computer World, No. 2, http://www.computerworld.com , (accessed May 2012)

Kernkamp, H.W.J., Van Dam, A., Stelling, G.S., De Goede, E.D., 2011. *Efficient scheme for the shallow water equations on unstructured grids with the application to the continental shelf.* Ocean Dynamics, 61, 1175–1188

Kramer S., Stelling G. S., 2008. *A conservative unstructured scheme for rapidly varied flows,* International Journal of Numerical methods for fluids, 58, 183–212.

Li, G., 2005. *Maintaining the Healthy Life of Yellow River,* the Yellow River Press, 203-245

Li, G., 2009. *The Water and Sediment Regulation on the Yellow River,* independently distributed keynote paper, Proceedings of the 4th International Yellow River Forum, Zhengzhou, China

Li, S., 2008. *Software System Requirements for Multi-reservoir-based Flood Control and Management - a case study for the Yellow River in China,* UNESCO-IHE MSc thesis (WSE-HI.08-04)

Li, S., Mynett, A. E., 2008. *Functional Requirements and Deployment Analysis for Multi-r,eservoir-based Flood Control and Management,* Advances in Water Resources and Hydraulic Engineering, 1 , 124-129

Li, S., Mynett, A. E., 2009a. *Function and Performance Requirements for Multi-reservoir-based Flood Control and Management Software System,* Proceedings of 8th International Conference on Hydroinformatics, Concepción, Chile

Li, S., Mynett, A. E., 2009b. *Computational Process Analysis for Multi-reservoir-based Flood Control and Management Software Systems,* Proceedings of the 33rd IAHR Congress, Vancouver, British Columbia, Canada,

Li, S., Mynett, A. E., 2010. *Traditional Methods and New Requirements of Reservoir Regulation Computation,* Proceedings of 9th International Conference on Hydroinformatics, HIC, Tianjin, China

Li, S., Mynett, A. E., 2011. *Towards Real-time Reactive Decision Support Systems for Flood Control and Flood Risk Management,* 5th International Conference on Flood Management (ICFM5), Tokyo-Japan

Li, S., Mynett, A. E., Popescu, I., 2012a. *On Delicate Reservoir Routing Processing,* Proceedings of 5th International Yellow River Forum, Zhengzhou China

Li, S., Mynett, A. E., Popescu, I., 2012b. *Overview of Reservoir Routing Methods,* Proceedings of 5th International Yellow River Forum, Zhengzhou China

Loucks, D. P., van Beek E., 2005. *Water Resources Systems Planning and Management - An Introduction to Methods, Models and Applications,* UNESCO Publishing, ISBN 92-3-103998-9, 342-354

Mann, R. I., Watson H., Cheney, P., Gallagher, C., 1989. *Accommodating Cognitive Style through DSS Hardware and Software,* Proceedings of 19th Annual Hawaii

International Conference on Systems Sciences, December 11-13, 1985. Republished in Decision Support Systems: Pulling Theory into Practice, Prentice-Hall, 118-130.

MikebyDHI, 2012. *Reservoir and Hydropower*, Mike Basin, Water Resources, Products. http://www.mikebydhi.com/Products/WaterResources/MIKEBASIN/ReservoirsHyd ropower.aspx (accessed Feb. 2012)

Millar, C.I., Stephenson, N. L., Stephens S.L., 2007. *Climate Change and Forest Change and Forests of the Future: Managing in the Face of Uncertainty*, Ecological Applications, 17(8), 2145-2151.

Milly, P. C. D., Betancourt, J.L., Falkenmark, M., Hirsch, R.H., Kindzewicz, Z., Lettenmaier, D.P., Stouffer, R.J., 2008. Stationarity is dead: Whither Water Management - Rethinking approaches to planning and design in a changing climate. Science, 319, 573-574.

Moya Quiroga, V., Popescu, I, Solomatine, D., L. Bociort, 2013. *Cloud and cluster computing in uncertainty analysis of integrated flood models*, Journal of Hydroinformatics, 15(1), 55-70

Moussa R., Bocquillon C., 2000. *Approximation Zones of De Saint Venant Equations for Flood Routing with Overbank Flow*, Hydrology and Earth System Sciences, 4(2), 251-261

Mynett, A.E., De Vriend, H.J., 2005. *Next Generation Information and Communication Technology (ICT) for Integrated Water Management of the Yellow River Basin*, Proceedings of the 2nd International Yellow River Forum

Mysiak J., Henrikson H. J., Sullivan C., Bromley J., Pahl-Wostl C., 2010. *The Adaptive Water Resource Management Handbook*, London, ISBN: 978-1-84407-792-2

National Research Council, 2004. Adaptive *Management for Water Resources Planning*, the National Academies Press, Washington, DC

National Flood Control and Drought Relief Command Headquarters (NFCDRCH), 2007. *The Design Book of National Flood Control System* (internal design document, in Chinese)- Sub-system for Office of National Headquarters

Nyberg, J.B., Taylor, B., 1995. *Applying adaptive management in British Columbia's forests* in Proceedings of the FAO/ECE/ILO International Forestry Seminar, Prince George, BC, 239-245

Nyberg, J.B., 1998. Statistics *and the practice of adaptive management, in Statistical Methods for Adaptive Management Studies*, V. Sit and B. Taylor, (Eds). Land Manage, Handbook 42, B.C. Ministry of Forests, Victoria, BC., 1-7

Official Journal of the European Union, 2007. *Directive 2007/60/EC of the European Parliament and of the Council of 23 October 2007 - on the assessment and management of flood risks*, http://eur-lex.europa.eu/LexUriServ/LexUriServ.do?uri= OJ:L:2007:288:0027:0034:EN:PDF (accessed Jan. 2012)

Popescu I., Jonoski A., van Andel S.J., Onyari E., Moya Quiroga V.G., 2010. *Integrated modelling for flood risk mitigation in Romania: case study of the Timis-Bega river basin*, International Journal of River Basin Management, 8, 269-280.

Popescu, I, Jonoski, A., Bociort, A., 2012. *Decision Support Systems for flood management in the Timis Bega catchment*, Environmental Engineering and Management Journal, 11(12), 2305-2311

Power, D.J., 2007. *A Brief History of Decision Support Systems*. DSSResources.COM, http://DSSResources.COM/history/dsshistory.html (accessed Sep. 2012)

Puls, L. G., 1928. *Construction of Flood Routing Curves*, House Document 185, U.S. 70th Congress, First Session, Washington, D.C.,

Raghunath, H. M., 2006. *Hydrology - Principles Analysis Design*, revised second edition, New Age International (P) Ltd., Publishers, 2006

Rist, L., Campbell, B., Frost, P., 2012. *Adaptive management: where are we now?* Environmental Conservation, 40 (1), 5-18

Rui X., 2004. *Principles of Hydrology* (in Chinese), ISBN 7-5084-2164-7 / TV · 496, China Water & Power Press, 228-229

Runge, M. C., 2011. *Adaptive management for threatened and endangered species.* Journal of Fish and Wildlife Management, 2(2), 220–233

Sammon, D., Carroll, E. 2001. *The Failure of a Decision Support System in Use: An Irish Case Study,* Americas Conferences on Information Systems (AMCIS) Proceedings

Senge, P. M., 1990. *The Fifth Discipline: The Art and Practice of the Learning Organization*, Currency Doubleday, New York.

Shu, Q., Deng, J., 2000. *Construction of Flood Control and Commanding System*, China Water Resources (in Chinese), Vol. 1

Smit B., Pilifosova, O., Burton, I. , Challenger, B., Huq, S., Klein, R. J. T., 2001. *Adaptation to Climate Change in the Context of Sustainable Development and Equity in McCarthy* J. et al, (Eds), Climate Change 2001: Impacts, Adaptation, and Vulnerability. (Working Group II of IPCC). Cambridge University Press. 879-885

Sprague, R. H., Carlson, E. D., 1982. *Building effective decision support systems*, Englewood Cliffs, Prentice-Hall, N.J

Stankey, G. H., Clark R. N., and Bormann B. T., 2005. *Adaptive Management of Natural Resources: Theory, Concepts, and Management Institutions*, United States Department of Agriculture, Forest Service, Pacific Northwest Research Station, General Technical Report, PNW-GTR-654

Stelling G. S., Verwey A., 2005. *Numerical Flood Simulation*, The Encyclopedia of Hydrological Sciences, JohnWiley & Sons Ltd

Stoker, J. J., 1957. *Water Waves, the Mathematical Theory with Application*, Interscience Press, New York

Szilágyi, J., Szöllösi-Nagy, A., 2010. *Recursive Streamflow Forecasting - A State-space Approach*, CRC Press, Taylor & Fancis Group, ISBN: 978-0-415-56901-9(Hbk), London, 12-13

Tang Y., Li, S., Tao, X., Yao, B., Zhang, F., Mynett, A. E., Xuan, Y., 2009. *Integrated Modeling of Flood Forecasting and Multi-reservoir-based Operation in Yellow River Basin*, China, Proceedings of the 4th International Yellow River Forum, Zhengzhou, China

Trenberth K., 2005. *The Impact of Climate Change and Variability on Heavy Precipitation, Floods, and Droughts*, Encyclopedia of Hydrological Sciences, M G Anderson (Eds)

UNESCO (Beijing Office), YRCC, 2011. Climate *Change Impacts and Adaptation Strategies in the Yellow River Basin*, Popular Science Press (China), Beijing. 272-230

Uran, O. , Janssen, R., 2003. *Why are spatial decision support systems not used? Some experiences form the Netherlands*, Computers, Environment and Urban Systems, 27 (5), 511-526

USACE website, 2011. *HEC-ResSim, Features*, http://www.hec.usace.army.mil/software/hec-ressim/ (accessed Dec. 2011)

USACE, 2007. HEC-ResSim Reservoir System Simulation, *User's Manual*, Version 3.0, 1-1,2; 13-1,3

van der Krogt, 2012. *3 Example RIBASIM7 projects*, PPT presentation

Van Delden, H., 2009. *Lessons learnt in the development, implementation and use of Integrated Spatial Decision Support Systems*, proceedings of the 18th World IMACS / MODSIM Congress, Cairns

van Leeuwen, E., Schuurmans, W., 2012. *Ten questions to... Professor Guss Stelling about 3Di water management.* Hydrolink, International Association for Hydro-Environment Engineering and Research, Hydrolink, 80-82

Van, P.D.T , Popescu, I., van Griensven, A., Solomatine, D., Trung, N. H., and Green, A., (2012), *A study of the climate change impacts on fluvial flood propagation in the Vietnamese Mekong Delta*, Hydrology and Earth System Sciences., 16, 4637–4649,

Verwey, A., 2005. *Hydroinformatics Support to Flood Forecasting and Flood Management*, Proceedings of the 4th Inter-Celtic Colloquium on Hydrology and Management of Water Resources, Guimarães, Portugal

Walters, C. J., Hilborn, R., 1978. *Ecological optimization and adaptive management*, Annual Review of Ecology and Systematics, 9,157 -188

Walters, C. J., 1986. *Adaptive Management of Renewable Resources*, New York, NY, USA: Macmillan.

Walters, C. J., Holling C.S., 1990. *Large-scale management experiments and learning by doing*, Ecology, 71, 2060- 2068.

Walters, C. J., 2007: Is adaptive management helping to solve fisheries problems? Ambio, 36, 304-307.

Wan, Z., 2002. *Functional Reactive Programming for Real-Time Reactive Systems*, a Dissertation Presented to the Faculty of the Graduate School of Yale University

Wang, G., Wu, B., Wang, Z., 2005. *Sedimentation problems and management strategies of Sanmenxia Reservoir*, Yellow River, China, Resources Researches, 41, W09417, DOI:10.1029/2004WR003919.

Wang, J., 2007. *Development of A Decision Support System for Flood Forecasting and Warning – A Case Study on The Maribyrnong River*, PhD thesis, Victoria University, Melbourne, Australia

Weber, L., 2007. *The need for science-based adaptive management for fish passage design. River Restoration: Practices and Concepts*, proceedings of Fish Passage on Midwestern Streams: Evaluation of Stability and Functionality of Dam Removals, Constructed Fishways and Culvert Crossings, Illinois Institute of Technology, Wheaton, IL

Werner, M., Van Dijk, M., Schellekens, J., 2004. Delft-FEWS: *An Open Shell Flood Forecasting System*, Proceedings of the 6th International Conference on Hydroinformatics - Liong, Phoon & Babovic (Eds), World Scientific Publishing Company, ISBN: 981-238-787-0

Westphal K. S., Vogel R. M., Kirshen P., Chapra S. C., 2003. *Decision Support System for Adaptive Water Supply Management*, Journal of Water Resources Planning and Management, 129 (3), 165-233

Williams, B. K., 2011. *Adaptive management of natural resources - framework and issues*, Journal of Environmental Management, 92, 1346-353

Williams, B. K., Brown, E. D., 2012. *Adaptive Management: The U.S. Department of the Interior Applications Guide. Adaptive Management Working Group*, U.S. Department of the Interior, Washington, DC. ISBN: 978-0-615-59913-7. p. v-vi

WMO-UNESCO, UNU, IAHS, 2007. *Concept Paper on International Flood Initiative*, P. 4-6, http://www.ifi-home.info/IFI_Concept_Paper.pdf (accessed Sep. 2012)

Wurbs, Ralph A., 2012. *Reservoir/River System Management Models*, Texas Water Resources Institute, Texas Water Journal, 3(1), 26–41

Xu, Z., Zhang, N., 2006. *On Precipitation Distribution and Change of the Yellow River Basin*, (China) Geographical Research (in Chinese), 25 (1), 27-34

Yan, G.J., 2004. *Research on Compensational Policy for Planned Flood Detention Operation in the Henan Section of the Lower Yellow River*, Yellow River (in Chinese), 26 (10)

Yellow River Conservancy Commission (YRCC), 2006a. *GIS-based Lower Yellow River 2D Flow and Sediment Transport Coupling Modeling*, Yellow River Conservancy Commission website, www.yrcc.gov.cn (accessed on June 2012)

Yellow River Conservancy Commission (YRCC), 2006b. *China National Flood Control and Hazard Relief Commanding System,* the Design Report of Yellow River Flood Control and Management Software System, Vol. No. NFCS-CMS-PD-03(2006)

Yellow River Conservancy Commission (YRCC), 2007. *China National Flood Control and Hazard Relief Commanding System*, the design Report of Yellow River Flood Control and Management Software System

Yellow River Conservancy Commission (YRCC), 2008. *Yellow River Flood Control Planning* (2008-2015/2025), Yellow River Process, China, 1, 57-58

Yellow River Conservancy Commission (YRCC), 2009. *Annual Flood Control and Management Strategy* (YRCC internal document)

Yellow River Conservancy Commission (YRCC), 2010. Annual *Flood Control and Management Strategy* (YRCC internal document)

Zhu, Q., Yu, Xin, Han, Q., Yang, M., Wang, Y., Liang G., 2005. *GIS-based 2D Flow and Sediment Transport Coupling Simulation of Lower Yellow River* (in Chinese), http://www.yellowriver.gov.cn/zlcp/kjcg/kjcg05_06/201108/t20110813_101725.html, (accessed Oct. 2012)

3Di Waterbeheer, 2013. *About 3Di Water Management*, http://www.3di.nu/home_english / (accessed Jan. 2013)

ACKNOWLEDGEMENTS

I would like to express my sincere appreciation and gratitude to all individuals and institutions who supported me during my PhD research and helped make my research become reality.

First of all, I would like to express my sincere thanks and appreciation to my supervisors, Prof. Arthur Mynett and Dr. Ioana Popescu. I am extremely grateful to Prof. Mynett, who stimulated me to complete my MSc degree and contiune with my PhD research. He initiated my research topic and has guided me throughout the entire research. Without his supervision I surely would not have succeeded. His expertise, support, and commitment were essential in all stages of my research. Both his broad advice on the direction of the research and his detailed technical instructions have given me the power and knowledge to do this research. I am also very grateful to my co-supervisor Dr. Ioana Popescu for all her guidance, suggestions and help on many subjects, as well as on solving numerous specific problems encountered during this research. Both supervisors have used all their expertise and wisdom to achieve the required quality of the research output. I am deeply grateful for all their effort and support.

During the research, very helpful suggestions and helps are obtained from experts in UNESCO-IHE and Deltares. I would like to express my sincere thanks to every one of them. The discussions we had always were very helpful and very much appreciated by me. I would like to express my sincere thanks to Mr. Adri Verwey, Dr. Micha Werner and Dr. Schalk Jan van Andel. They have given me very helpful suggestions, directions and references at the early stage of my research. Sincere thanks are also due to Dr. Andreja Jonoski, Dr. Hans Goossens and Dr. Zhengbing Wang for giving me suggestions on my research. My thanks also go to Ir. Karel Heynert and Dr. Dirk Schwanenberg for sharing their experience with me.

Kind help and support was also obtained from numerous staff members at UNESCO-IHE and Deltares. I would like express my sincere thanks to Ms Jolanda Boots, Ms Sylvia van Opdorp-Stijlen, Ms Tonneke Morgenstond-Geerts, Ms Martine Roebroeks-Nahon, Ms Anique Alaoui-Karsten, Mr Peter Stroo, Ms Shirley Dofferhoff, Ms Frances Kelly, Mr Erick de Jong, Ms Gretchen Gettel, Ms Esther de Groot, Mr. Edward Melger.

My special gratitude goes to Mr. Li Guoying, vice Minister of the Ministry of Water Resources and former Commissioner of the Yellow River Conservancy Commission (YRCC), Ministry of Water Resources, China, for his continuous support and kind care during my research. I also express my gratitude to Mr. Chen Xiaojiang, the present Commissioner of YRCC, for his kind support on my future career. The same applies to Mr. Chen Lvping, deputy director of the YRCC department of Personnel for his continued care and support. My thanks also go to Mr. Shan Hongqi, Ms. Sun Feng of the department of International Cooperation, Science and Technology, and to Mr. Lou Yuanqing and Mr. Li Yinquan of the Information Centre of YRCC.

I would also like to express my special thanks to Mr. Liao Yiwei, the Vice Commissioner of YRCC, Mr. Zhang Jinliang, general manager of the design institute of YRCC, for all their support. Sincere thanks go to Mr. Bi Dongsheng, Mr. Zhang Yong, Mr. Wang Zhenyu, Mr. Li Yuelun, Ms. Liu Xiaoyan, Mr. Wei Jun, Mr. Wei Xiangyang and all other colleagues in the Division of Flood Control Scheme and ICT Applications of the Flood Control Office, YRCC, for their support and for the extra burden they had to bear when I was absent from my office.

Many friends in Delft have also contributed to my research life in the Netherlands, in one way or the other; they will be remembered for the rest of my life: Ms. Tang Yiyuan, Mr. Mario Castro Gama, Mr. Xu Zheng, Mr. Ye Qinghua, Ms. Wan Taoping, Mr. Chen Hui, Ms. Zhu Xuan, Ms. Lin Yuqing, Mr. Yan Kun, Mr. Chen Xiuhan, Ms. Sun Wen, Mr. Guo Leicheng, Ms. Li Hong, Mr. Xu Zhuo, Ms. Qi Hui, Xu Min, Mr. Wang Chunqing, Mr. Yang Zhi, Mr. Cui Haiyang, Ms. Zhang Congjiao, Ms. Li Zhiping, Mr. Li Yanbin, Mr. Dong Wei, Mr. Chu Jing, Mr. Li Shanfei, Mr. Wang Wenbo, Ms. Duong and Ms. Hoang.

I am very grateful for the financial support of Deltares for enabling my PhD research and I give my grateful thanks to UNESCO-IHE for providing me with such excellent and stimulating environment and conditions to complete this research.

Finally, I would like to give my special thanks and express my heartfelt appreciation to my wife Liu Yuwen and my daughter Li Xinyun for their unconditional love and full support. Though they had to overcome lots of difficulties when I was not home, they had constantly showed me their love and continuously encouraged me during the research.

Shengyang LI

UNESCO-IHE Institute for Water Education,

Delft, the Netherlands

April 2013

ABOUT THE AUTHOR

Shengyang LI was born in Linshu, Shandong Province, China on 9 June 1964. He studied hydrology and water resources management in Chengdu University of Science and Technology (now Sichuan University) from 1983 to 1987 and graduated with a Bachelor of Engineering degree. He then worked at the Computerized Flood Forecasting Center (CFFC) of the Yellow River Conservancy Commission (YRCC) of the Ministry of Water Resources of China, as software engineer on hydrological software application development. He became the director of the Software Development Office of CFFC in 1994. In 1997, he worked on software system design, development and network management in the Information Center of YRCC and became the deputy director of the Network Center of the Information Center. From 1999 to 2001, he qualified as senior engineer and was appointed director of the Computer Training Center of the Information Center, YRCC. In 2002, he was appointed as the deputy director of the Data Management Center of the Information Center working on data center management.

In 2002, he joined the first young professionals training group of YRCC and began the Master's program at UNESCO-IHE. In 2003 he graduated with the Master of Engineering diploma and although he qualified for the Master of Science program, was called back to China by the YRCC to deal with some emergency flood issues. Upon his return in September 2003 he was assigned to the Flood Control Office of the YRCC headquarters to work as the deputy director of the division of Flood Control Schemes and ICT Applications. His main responsibility was to develop real-time flood control scenarios for flood management and decision support. In 2007 he returned to Delft to continue his MSc study on Hydroinformatics at UNESCO-IHE, sponsored by Deltares, under the supervision of Prof. Mynett. In April 2008 Mr. Shengyang Li obtained his Master of Science degree (with distinction) and continued to pursue his PhD at UNESCO-IHE.

He joined several large software development projects at YRCC, including the Digital Yellow River Project, the Yellow River Flood Control Software System of the national flood control commanding system, and the national ninth five-year planning project '3S Application' for the lower Yellow River flood monitoring. He is presently senior engineer in flood control management and ICT applications.

He married with Liu Yuwen in 1991 and got a daughter, Li Xinyun in 1995.

PUBLICATIONS

Li, S., Mynett, A. E., Popescu, I., 2013. *A Flexible Reservoir Routing Method*. International Journal of Hydrological Processes (in preparation)

Li, S., Mynett, A. E., Popescu, I., 2013. *Real-time Reactive Mechanism for Decision Support Systems*. International Journal of Environmental Modeling and Software (in preparation)

Li, S., Mynett, A. E., Popescu, I., 2013. *Multi-reservoir Based Flood Control Decision Support System for the Yellow River*. International Journal of Environmental Modeling and Software (in preparation).

Li, S., Mynett, A. E., Popescu, I., 2013. *Multi-reservoir Based Flow Regulation - an approach towards complex uncontrolled routing process*. International Journal of Hydrological Processes (in preparation)

Castro-Gama, M., Popescu, I., Li, S., Mynett, A. E., Van Dam, A., 2013. *Statistical flood inference model based on the simulation of a multiple reservoir system: the case study of the Yellow River, China*. Environmental Modeling and Software. (Under review)

Castro-Gama, M., Popescu, I., Li, S., Mynett, A. E., Van Dam, A., 2013. *New modeling approach for flooding events: the case study of the Yellow River in China*. Environmental Modeling and Software. (Under review).

Li, S., Mynett, A. E., Popescu, I., 2012. *Adaptive DSS-oriented Computational Core for Multi-reservoir-based Flood Control and Water Resources Management*. 5th International Yellow River Forum, Zhengzhou, China

Li, S., Mynett, A. E., Popescu, I., 2012. *Delicate Computations for Complex Release Schemes of Reservoir-based Flood Control and Water Resources Management*. 5th International Yellow River Forum, Zhengzhou, China

Li, S., Mynett, A. E., 2011. *Towards Real-time Reactive Decision Support Systems for Flood Control and Flood Risk Management*. 5th International Conference on Flood Management, Tokyo Japan

Li, S., Mynett, A. E., 2010. *Traditional Methods and New Requirements of Reservoir Regulation Computation*. 9th International Conference on Hydroinformatics, Tianjing, China

Li, S., Mynett, A. E., 2010. *Adaptive Management of Multi-Reservoir River Systems*. *9th International Conference on Hydroinformatics*, Tianjing, China

Mynett, A. E., Li, S., 2010. Adaptive Management of Multi-Reservoir River Systems – a Case Study for the Yellow River. 1st European IAHR division congress, UK

Li, S., Mynett, A. E., 2009. *On the information and functional requirements of the software system for real-time comprehensive flood control meeting and decision support*. 4th International Yellow River Forum, Zhengzhou, China

Li, S., Mynett, A. E., 2009. *The dialectic and philosophical analysis of several relationships in flood control decision.* 4th International Yellow River Forum, Zhengzhou, China

Li, S., Mynett, A. E., 2009. *Function and Performance Requirements For Multi-Reservoir-Based Flood Control And Management Software System.* 7th International Symposium on Ecohydraulics and 8th International Conference on Hydroinformatics, Concepción, Chile.

Tang, Y., Li, S., Tao, X., Yao, B., Zhang, F., Mynett, A. E., Xuan, Y., 2009. *Integrated Modeling of Flood Forecasting and Multi-reservoir-based Operation in Yellow River Basin,* China. 4th International Yellow River Forum, Zhengzhou, China

Li, S., Mynett, A. E., 2008. *Functional Requirements and Deployment Analysis for Multi-Reservoir-Based Flood Control and Management Software System.* Joint 16th IAHR-APD and 3rd IAHR-ISHS Conference, Nanjing China.

Li, S., 2007. *Analysis on Potential Measures to Deal with the Silting of the Xiaolangdi Reservoir, Developments and Equipments in Reservoir Sedimentation Management* (in Chinese), YRCC (Eds), Yellow River Press, ISBN: 978-7-80734-215-1

Li, S., 2003. *Data Storing and Management Analysis for the Digital Yellow River Project* (chapters 2, 4), Digital Yellow River Planning (in Chinese), YRCC (Eds), ISBN: 7-80621-714-2/TV•328

Wu, H., Li, S., Teng, Y., 2001. *Data Retrieving System for Yellow River Tele-metering Data.* Proceedings of OA' 2001- International Conference on Office Automation, (China) E-Industrial Press, ISBN: 7-5053-5055-2/Z.336

Li, S., Wang, D., Zhao, Y., 2001. *3S (GIS, GPS, RS)-based Real-time Monitoring of Flood and Inundation on Lower Yellow River.* Proceedings of OA' 2001- International Conference on Office Automation, (China) E-Industrial Press, ISBN: 7-5053-5055-2/Z.336

Li, S., 1998. *Three-Reservoir based Flood Control Regulation* (Chapter 5). The Yellow River Flood and Ice Control Decision Support System (in Chinese), Cui, J., Cai, L., Xin, G., (Eds), Yellow River Press, ISBN: 7-80621-144-6/TV-105

T - #0411 - 101024 - C210 - 240/165/11 - PB - 9781138001022 - Gloss Lamination